Solid State
Surface Science

VOLUME 3

SOLID STATE
SURFACE SCIENCE

Edited by Mino Green

Department of Electrical Engineering
Imperial College of Science and Technology
London, England

VOLUME 3

MARCEL DEKKER   *New York*   *1973*

LIBRARY OF CONGRESS CATALOG CARD NUMBER: 77-85241

ISBN: 0-8247-6017-4

PRINTED IN THE UNITED STATES OF AMERICA

sd
11/24/73

To my mother
Elizabeth Green

PREFACE

Science has been defined as the acquiring of knowledge and the construction of conceptual models of the natural world. Accepting this definition I hope the reader will agree that the three chapters which go to make up this book are interesting examples of science at various stages of data acquisition and model building.

The first chapter, by Logan, concerns the theory of gas-surface scattering and accommodation, which for all its past history is a rapidly developing field. A detailed discussion is given of what happens when an atom or molecule collides with the surface of a solid. Classical and quantum mechanical models are discussed. The kind of useful information obtainable from molecular beam experiments, which is not only relevant to the motion of earth satellites but also to surface chemical reactions on terra firma, is fully discussed.

The second chapter, by Frankl, is a re-examination of the theory of surface space-charge layers in solids. The underlying assumptions and the nature of the approximations are discussed with a view to pointing the way to useful extensions of the theory. Emphasis is placed on quantization effects and space-charge capacitance. Materials ranging from near-insulators to semimetals are covered in a unified manner. Metals are excluded from the discussion because of the overwhelming influence of many-body effects.

The final chapter is about one of the newer methods of surface investigation, namely field emission spectroscopy of surfaces having chemisorbed atoms. The authors, Gadzuk and Plummer, are mainly concerned with the important information obtainable from the energy distribution of field emitted electrons. A theoretical introduction is followed by a review of recent results.

I hope that the reader will show some indulgence toward the mixture of units which are in use in this series. Some scientists have embraced S.I.

units while others are still fighting the old battle of cgs versus mks. One can only regret that science has built its very own tower of Babel.

Again I take much pleasure in thanking my wife, Diana, and my secretary, Miss Simone Ruddy, for invaluable help and encouragement in connection with editing this series.

LIST OF CONTRIBUTORS

DANIEL R. FRANKL, Department of Physics, The Pennsylvania State University, University Park, Pennsylvania

J. W. GADZUK, National Bureau of Standards, Washington, D.C.

R. M. LOGAN, Royal Radar Establishment, Malvern, Worcestershire, England

E. W. PLUMMER, National Bureau of Standards, Washington, D. C. *

*Present address: Department of Physics, University of Pennsylvania, Philadelphia, Pennsylvania.

CONTENTS OF VOLUME 3

CONTENTS OF OTHER VOLUMES

Solid State
Surface Science

VOLUME 3

CHAPTER 1

THEORY OF GAS–SURFACE SCATTERING AND ACCOMMODATION

R. M. Logan

Royal Radar Establishment
Malvern, Worcestershire
England

1

1. INTRODUCTION

This chapter deals with theoretical studies of the collision between a gas atom (or molecule) and the surface of a solid. The details of this interaction have traditionally been expressed in terms of various accommodation coefficients, such as the energy accommodation coefficient. Such accommodation coefficients are not adequate, however, to describe the full details of the gas-surface scattering process. For example, in modern molecular beam scattering experiments, the full flux and velocity distributions of the scattered atoms can be measured, and the present trend is to concentrate on these detailed distributions rather than overall accommodation coefficients.

The increased interest in the subject of gas-surface interactions over the last ten years has been largely due to its relevance to the motion of bodies such as earth satellites under rarefied gas conditions, that is to say conditions in which the mean free path in the gas is large compared to the characteristic length scale of the body. The interaction of a body with such a rarefied gas is governed by the nature of the individual collisions between the gas molecules and the surface, rather than by boundary layer or bulk flow effects.

In addition to applications related to rarefied gas dynamics there are other more general, and in the long run perhaps more important, reasons for studying gas-surface collisions. Many processes such as surface chemical reactions, catalysis and growth of crystals from the vapor involve the adsorption of the gas species at the surface, and in these cases it is important to know whether the gas molecule will be trapped at the surface or reflected on its initial encounter with the surface. The calculation of these trapping probabilities, and also the finer details of the interaction, is one of the aims of gas-surface collision theory. At present, calculations of trapping probabilities have only been made for very simple systems, such as inert gases on clean surfaces, and even these calculations are not very reliable.

Another important reason for studying gas-surface collisions lies in the fact that the ability to understand and interpret the scattering of atoms from surfaces would provide one with a powerful tool for the study of the surfaces

themselves. An atom incident with thermal energy on a solid does not penetrate beyond the outer atomic layer of the solid, and this represents an important advantage over a technique such as LEED or HEED, where even at
the lowest usable energies the electrons penetrate to some extent into the
solid. Thus the scattering of an atomic beam of an inert gas with thermal
energy should provide information about the topography, structure and degree of impurity coverage of the surface.

Unfortunately there is no single unified theory of gas-surface collisions
which deals in a manageable way with all the features of interest. In fact,
there is a great abundance of theories in current use, ranging from very
simple to complicated, some using classical mechanics and some using
quantum mechanics. This vast array of theories is a source of some con-
fusion to people working in the field, not to mention outsiders. Some of
these theories are approximately valid, or at least useful, in certain re-
gimes; some of them are not.

In Section 2 we give a brief introduction to some of the concepts and ter-
minology used in the study of gas-surface collisions, and discuss some of
the general features which are common to most of the theoretical treatments.
We also give in Section 2 a brief outline of the experimental background to the
problem. Section 3 comprises a survey of the various theories, with a view
to indicating the essential features of the different theories, discussing their
likely validity and stating what useful results, if any, they have led to. In
this survey we concentrate on the physical content of the theories, rather
than discuss in depth the details of the analyses involved. Finally in Section 4 we consider trapping effects and discuss briefly the various theoretical attempts to calculate the trapping probability in simple systems.

2. GENERAL FEATURES OF GAS-SURFACE COLLISION THEORIES AND EXPERIMENTAL BACKGROUND

Before considering individual theoretical treatments in Section 3 we consider various definitions and concepts which are common to all the treatments. A typical gas surface scattering event is illustrated in Fig. 2.1.

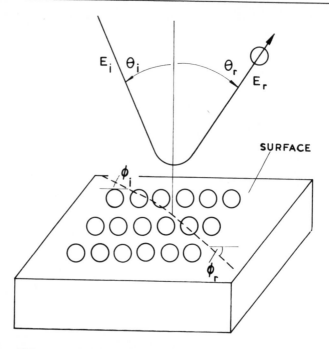

FIG. 2.1. Representative gas-surface scattering event.

The atom approaches the surface in a given direction and with a given energy, interacts with the solid and, if it is not trapped, departs with some final direction and energy. A simplification which is universally used in this subject is to assume a binary additive interaction potential between the gas atom and each solid atom. Thus the overall interaction potential is given by

$$V(\underline{r}, \underline{R}_1, \ldots) = \sum_{i}^{N} V_i\left(\left|\underline{r} - \underline{R}_i\right|\right) \qquad (2.1)$$

where V_i is the binary interaction potential between the gas atom at $\underline{r}$ and the ith solid atom at $\underline{R}_i$. For the binary potential V_i a convenient analytical form such as the Lennard-Jones or Morse potential is usually chosen. The actual expressions for these and alternative potentials are given below in the discussion of specific theories. In writing Eq. (2.1) we are implicitly assuming the Born-Oppenheimer approximation [1], which means that the electrons which give rise to the interaction potential do not otherwise enter into the dynamics of the interaction. For a comprehensive treatment of binary potentials see Hirschfelder, Curtiss, and Bird [2]. Gas–solid interaction potentials, and the assumption of additivity of the potentials as used in Eq. (2.1), are considered by Young and Crowell [3], Kaminsky [4], de Boer [5], and Beder [6]. The general conclusion from these studies is that the above assumptions are reasonable for gas atoms with energies up to about a few electron volts interacting with nonmetals. For metals the principal interaction is between the gas atom and the conduction electrons [3]. Thus an expression such as Eq. (2.1) cannot strictly apply. One consequence of this interaction with the conduction electrons is that the periodic nature of the gas–surface potential (along the surface) is much weaker than for ionic or covalent solids.

Another feature which is common to most of the theories to be discussed is the "perturbation" nature of the analysis used. This is perhaps not surprising in view of the obvious many-body character of the gas–surface collision problem. In many of the cases where the perturbation treatment has been used, it is probably not valid. The usual procedure is to find a zero-order solution to the equation of motion of the gas atom, keeping the atoms of the solid fixed at rest. In classical mechanics the result is a trajectory $\underline{r}(t)$ for the gas atom. Corresponding to this trajectory there is a zero-order force pulse given by $F = -\nabla V$, where the potential V is evaluated with the surface atoms rigidly fixed. With this force pulse $F(t)$ the energy transferred to the solid can be calculated, and this is taken to be equal to the energy lost by the gas atom. A similar procedure applies in quantum mechanics when we deal with a zero-order set of states, rather than a classical trajectory. The methods are discussed in more detail below in connection with specific theories. This perturbation character of many of the gas–surface collision theories has been discussed by Gilbey [7] and Goodman [8]. There are types of theoretical treatment which do not rely on the perturbation technique; these can be roughly divided into three classes: (1) those which assume an impulsive (hard) interaction between the gas and surface atoms; (2) those which use iterative computer calculations, or numerical integration of the classical equations of motion, (3) recent quantum

mechanical treatments (see Section 3.1.3) which express the required transition probabilities in terms of a "perturbation series," and enable exact results to be obtained at least in principle.

There are two terms which occur frequently in discussion of the theories
which are conveniently defined at this stage, namely the energy accommodation coefficient and the scattering distribution. The energy (or thermal) accommodation coefficient α is defined as follows [9]:

$$\alpha = \frac{\overline{E}_r - \overline{E}_i}{\overline{E}_s - \overline{E}_i} \tag{2.2}$$

where $\overline{E}_i$ is the mean energy per atom incident on the surface, $\overline{E}_r$ is the actual mean energy per atom leaving the surface, and $\overline{E}_s$ is the mean energy
per atom leaving the surface in the limiting case when the gas has come into thermal equilibrium with the surface. The quantity α is therefore a
measure of the efficiency with which the gas comes into equilibrium with
the surface. For monatomic gases the energy referred to above is, of
course, translational energy; for polyatomic gases other energy accommodation coefficients referring specifically to energy associated with internal rotational or vibrational modes can be defined. Calculations involving
these internal modes are considerably more difficult than those involving
only translational energy, and relatively little progress has been made in
this area. In this review we are almost exclusively concerned with monatomic gases.

The concept of an accommodation coefficient as given by Eq. (2.2) is
only really useful when dealing with a 3-dimensional isotropic gas. For
cases involving the scattering of molecular beams, where the incident and
reflected molecules have distinct directional properties, the averaging over
angular direction as implied in Eq. (2.2) obviously obscures important features of the scattering. It is possible to define a reduced accommodation
coefficient which is a function of the reflected angle, but such definitions
have not proved very useful. It is far more useful to think in terms of the
complete velocity distribution function of the scattered atoms $f_+(\underline{v})$. In
practice most of the experimental measurements and theoretical calculations are made in the plane defined by the incident beam and the normal to

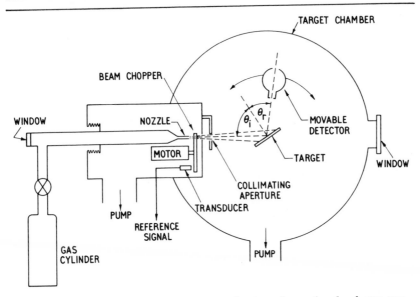

FIG. 2.2. Representative experimental set-up for molecular beam surface scattering; based on that of Yamamoto and Stickney [10].

the surface, and in this plane the reflected atoms can be described in terms
of the outgoing angle θ_r (see Fig. 2.1) and their speed at this angle. In this
case, rather than the complete distribution function $f_+(\underline{v})$, it is convenient to
work with the quantities $F(\theta_r)d\theta_r$, which gives the flux of atoms scattered into the angular range θ_r to $\theta_r + d\theta_r$ (assuming a fixed element of solid angle),
and $f_\theta(v)$ which is the speed distribution function of the atoms scattered at the
angle θ_r.

We next consider briefly some of the experimental features of the gas-
surface interaction which are necessary background for the present review.
A typical experimental set-up for measurements on molecular beam surface
scattering is shown in Fig. 2.2, adapted from Ref. [10]. The angles θ_i and
θ_r shown in this figure are measured in the plane containing the incident
beam and the surface normal, as discussed in the previous paragraph. The
apparatus illustrated uses an aerodynamic nozzle source to generate the
beam, which gives an approximately monoenergetic beam; this is advantageous from the point of view of interpreting the experimental results. Many of

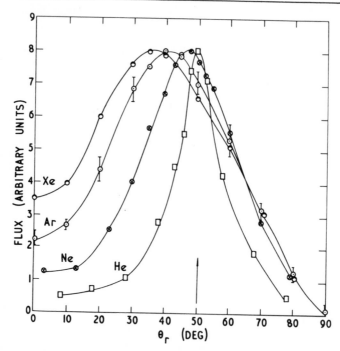

FIG. 2.3. Experimental scattering distributions for various gases scattered from Ag(111). $\theta_i = 50°$, $T_s = 560°K$, $T_g = 300°K$ (Saltsburg and Smith [90]).

the experiments, particularly the earlier ones, have used a Knudsen oven beam source which gives a Maxwellian velocity distribution in the beam at a temperature corresponding to that of the oven.

Some characteristic scattering patterns obtained from an experimental arrangement such as that in Fig. 2.2 are shown in Figs. 2.3 and 2.4, showing the flux of scattered atoms as a function of the scattering angle θ_r. These results demonstrate two features which are observed fairly generally. From Fig. 2.3 we see that the width and position of the lobular pattern varies with different gas species, for given surface and temperature conditions. This change occurs mainly due to the changing atomic mass of the gas atom, and changing interaction potential with the surface. Fig. 2.4 illustrates the fact

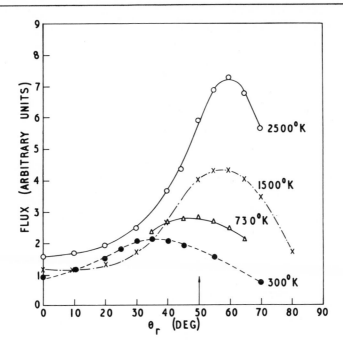

FIG. 2.4. Experimental scattering distribution for Xe/Ag(111) at various T_g. $\theta_i = 50°$, $T_s = 560°$ K (Saltsburg and Smith [90]).

that as the surface-temperature/gas-temperature ratio increases the peak of the scattering lobe shifts towards the surface normal.

In fact a wide variety of behavior is observed in gas-surface scattering experiments, which can be roughly divided into three categories:

(1) Quantum mechanical diffraction effects; these manifest themselves as multiple peaks in the scattering pattern, and are only observed with the lighter atoms (e.g., He) whose de Broglie wavelength is comparable with the lattice spacing of the solid. Theoretical studies associated with these diffraction patterns are considered in Section 3.1.

(2) Thermal scattering. The examples shown in Figs. 2.3 and 2.4 fall into this category. In this case the scattering behavior (i.e., width and

position of the lobe) is significantly affected by the surface temperature.
For example, increasing the surface temperature causes the scattering
lobe to move towards the normal and broaden. This type of behavior is
obtained with beam energies up to a few tenths of an eV. Theoretical
studies associated with thermal scattering are considered mainly in Section 3.4.

(3) Structure scattering. This is found at high beam energies (of the order of an eV and above) when the thermal energy of the solid is negligible
compared with the translational energy of the beam. The scattering behavior is now determined by the structure of the gas-surface interaction
potential, which is itself a function of the incident energy of the gas atom.
At high enough energies the scattering approximates that from an array
of hard-spheres. The scattering patterns in structure scattering may
exhibit multiple peaks, although the cause of these is quite different from
the diffraction phenomena mentioned above. Theoretical studies associated with structure scattering are considered mainly in Section 3.5.

We do not attempt here to give a review of the available molecular beam
scattering results. A review of the data published up to about 1967 has been
given by Stickney [11]. Many important experimental results have been published since then, such as those given in Refs. [12]-[22].

Finally we consider the experimental measurement of the energy accommodation coefficient. Reviews of the experimental procedure and typical results are given by Thomas [23] and Wachman [24]. Two important methods
for measuring energy accommodation coefficients are Knudsen's low pressure (LP) method [25] and the temperature jump (TJ) method [24]. In both
methods the cell containing the experimental gas consists of a large-diam-
eter cylindrical tube containing an axially mounted filament wire of small
diameter. The filament is maintained electrically at a temperature somewhat higher than the tube walls, which are surrounded by a constant-tem-
perature bath. Experimental measurements are made of the power loss per
unit area from the filament to the gas W_g, the gas pressure, filament
temperature T_S, and the tube wall temperature T_W.

Using the LP method the accommodation coefficient α is given by W_g/W^*, where W^* is the kinetic theory value for the energy transfer from the surface assuming a gas arriving at the filament at temperature T_g and leaving at T_s (i.e., complete accommodation). With the LP method the mean free path is sufficiently large for the assumption $T_g = T_W$ to be valid. With the TJ method higher pressures in the tube are used and the assumption $T_g = T_W$ is not valid. The interpretation of the TJ data is more involved than the interpretation of the LP data, and the theoretical basis is not so well founded, being based on the kinetic-theory relation between the temperature jump distance and the accommodation coefficient; for further details see Refs. [24] and [26]. The various different theoretical approaches to the calculation of the energy accommodation coefficient are described in Section 3.

3. SPECIFIC GAS-SURFACE COLLISION THEORIES

3.1. QUANTUM TREATMENTS

Several distinct quantum mechanical treatments of the gas-surface collision problem have been presented, starting with the work in the 1930's which is usually associated with Lennard-Jones and Devonshire (as discussed below). A general review of the quantum mechanical treatments up to about 1966 has been given by Beder [6]. Recently Goodman [27] has given a critical review of the work of Lennard-Jones and co-workers [28-37] and put forward several important extensions of this work. We begin this section with a brief outline of the Lennard-Jones and Devonshire theory of inelastic scattering; although this theory is somewhat dated and has not provided useful agreement with experiment, it deserves consideration because it has served as the basis for many other quantum mechanical treatments. It also provides a convenient framework in which to discuss some current developments. The question of the validity of the Lennard-Jones and Devonshire theory and a discussion of the recent developments is given in Section 3.1.2. In Section 3.1.3 we take up the problem of elastic scattering (diffraction), that is to say processes in which no phonons are annihilated or created in the solid.

3.1.1. Lennard-Jones and Devonshire Theory of Inelastic Scattering

We consider here the case of inelastic scattering, that is the case where energy exchange takes place between the gas atom and the solid. In particular we consider the calculation of the energy accommodation coefficient. The first steps in this direction were given in the work of Zener [38-40] and of Jackson, Howarth, and Mott [41-43]. In the treatment of Jackson and Mott [42] the gas atom interacts with a single-frequency oscillator (i.e., an Einstein lattice) through a purely repulsive potential. This work was extended soon afterwards by Lennard-Jones and Devonshire [28-32], Devonshire [33, 34], and Lennard-Jones and Strachan [35-37], who replaced the interaction potential by a Morse potential and used a Debye frequency spectrum for the normal modes of the lattice. This model, usually associated with the names of Lennard-Jones and Devonshire (hereafter referred to as LJD), has been used as a basis for many other treatments up to quite recent times, but as we shall see below has not produced any useful results as far as inelastic scattering is concerned and is probably not valid.

The situation analyzed by LJD is illustrated in Fig. 3.1. The gas atom of mass m_g approaches an idealized lattice with kinetic energy E_i. Only the motion of the gas atom perpendicular to the solid is considered, and the gas atom is assumed to have a head-on interaction with a single surface atom of mass m_S. We denote the coordinates of the gas atom and the target atom in the direction normal to the surface by z_g and z_S, respectively. The Morse potential function is then given by

$$V = D \left\{ \exp \left[-2\kappa (z_g - z_S) \right] - 2 \exp \left[-\kappa (z_g - z_S) \right] \right\}, \tag{3.1}$$

where D is the well depth and κ is an inverse length parameter describing the hardness of the interaction. With the perturbation treatment in mind the first step is to consider the interaction with the target atom fixed in its equilibrium position $z_S = z_{S0}$. The Schrödinger equation for the wave function ψ describing the gas atom is

$$-\frac{\hbar^2}{2m_g} \frac{d^2\psi}{dz_g^2} + V(z_g, z_S)\psi = E\psi \tag{3.2}$$

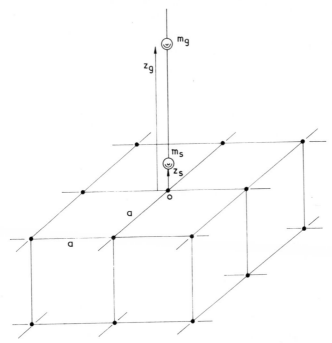

FIG. 3.1.   Gas-surface collision as analyzed by Lennard-Jones and Devonshire.

The solution of Eq. (3.2) obeying the required boundary conditions is given in Ref. [34] and will not be reproduced here.

The perturbing potential in the present problem is obtained by keeping the first-order term in the expansion of the potential:

$$H' \equiv V(z_g, z_s) - V(z_g, z_{s0}) \simeq (z_s - z_{s0}) V'(z_g) \tag{3.3}$$

where

$$V'(z_g) = [dV(z_g, z_s)/dz_s]; z_s = z_{s0} \tag{3.4}$$

We consider the transition from the state $|E_i, n_i>$ to the state $|E_f, n_f>$ where E_i and E_f are the initial and final states of the gas atom, and n_i and

n_f are the initial and final states of the qth mode of the lattice. The matrix element on which the transition probability depends factors as follows:

$$<E_f, \; n_f | H' | E_i, \; n_i> \; = \; <n_f | z_s - z_{s0} | n_i> <E_f | V'(z_g) | E_i> \tag{3.5}$$

The lattice displacement matrix is given by [6]

$$| <n_f | z_s - z_{s0} | n_i> |^2 \; = \; \frac{1}{Nm_s} \frac{\hbar}{2\omega_q} \left(n_q + \frac{1}{2} \pm \frac{1}{2} \right) \tag{3.6}$$

where n_q is the occupation number of the qth mode, ω_q is its frequency, and N is the total number of atoms in the crystal. The plus and minus signs apply for the creation and annihilation, respectively, of a phonon of frequency ω_q. The fact that the expression allows only single-phonon transitions is a direct consequence of keeping just the first-order term in Eq. (3.3). The gas atom matrix in Eq. (3.5) is given by [34]

$$|<E_f | V' | E_i>|^2 \; = \frac{1}{4} h^2 \kappa^2 \left(p_f^2 - p_i^2 \right) \frac{\sinh 2\pi p_f \; \sinh 2\pi p_i}{(\cosh 2\pi p_f - \cosh 2\pi p_i)^2} \frac{\left(A_i + A_f \right)^2}{A_i A_f} \tag{3.7}$$

where $p_f^2 = 2m_g E_f / \kappa^2 \hbar^2$ (similarly p_i),

$$A_f \equiv | \Gamma \left(-d + \frac{1}{2} + ip_f \right) |^2 \tag{3.8}$$

(similarly A_i), and $d^2 \equiv 2m_g D / \kappa^2 \hbar^2$. The quantity p corresponds to the gas atom momentum, and the significance of the quantity d is that the number of bound states of the Morse potential, Eq. (3.1), is given by the integer closest to d.

We denote by $P(E_i; n_f, n_i)$ the probability per collision that a gas atom with initial energy E_i changes the mode from state n_i to n_f. Using first-order time-dependent perturbation theory this is given by

$$P(E_i; \; n_f, \; n_i) \; = \frac{2\pi}{\hbar} \frac{\rho (E_f)}{N(E_i)} |<E_f, \; n_f | H' | E_i, \; n_i>|^2 \tag{3.9}$$

where $\rho(E_f)$ is the density of states for the gas atom and $N(E_i)$ is the collision rate of incident gas atoms. The average energy transfer to the mode q of the lattice is then given by

$$\overline{\Delta E}_q = \hbar\omega_q \int_0^\infty f(E_i) \, dE_i \left[\sum_{n_q = 1}^\infty w(n_q) \, P(E_i; n_q, n_q - 1) \right.$$

$$\left. - \sum_{n_q = 0}^\infty w(n_q) \, P(E_i; n_q, n_q + 1) \right] \qquad (3.10)$$

where $f(E_i)$ is the distribution function for the energy of the incident atoms, given by Maxwellian distribution

$$f(E_i) = (kT_g)^{-1} \exp(-E_i/kT_g) \qquad (3.11)$$

$w(n_q)$ is the probability of finding the qth mode with occupation number n_q, given by

$$w(n_q) = \left(1 - \exp\left(\frac{-\hbar\omega_q}{kT_s}\right)\right) \exp\left(\frac{-n_q \hbar\omega_q}{kT_s}\right) \qquad (3.12)$$

In the above expressions T_g and T_s refer to the temperatures of the gas and solid, respectively. The transition probabilities have the following symmetry property (corresponding to microscopic reversibility)

$$P(E_i; n_q + 1, n_q) = P(E_i - \hbar\omega_q; n_q, n_q + 1) \qquad (3.13)$$

In this analysis trapping of the gas atoms at the surface is neglected, so we put

$$P(E_i; n_q + 1, n_q) = 0 \qquad (3.14)$$

if $E_i < \hbar\omega_q$.

The general definition of the energy accommodation coefficient is given in Eq. (2.2). LJD [34] and many others use a slightly less general definition by taking the limit $T_s = T_g$. We denote this coefficient by α_{lim} to distinguish it from the more general one. Thus,

$$\alpha_{lim} = \left[\lim_{T_g \to T_s \to T} \right] \frac{\overline{E}_f - \overline{E}_i}{\overline{E}_s - \overline{E}_i} \qquad (3.15)$$

Summing Eq. (3.10) over the normal modes of the crystal (assuming a
Debye frequency spectrum) and using Eqs. (3.13) and (3.14) we obtain the
following expression for the accommodation coefficient

$$\alpha_{lim} = \lim_{T_g \to T_s \to T} \sum_q \frac{\overline{\Delta E_q}}{k(T_s - T_g)}$$

$$= \frac{3(2\pi)^3}{\omega_m^3} \int_0^{\omega_m} \frac{(\hbar\omega/kT)^3 \omega^2 d\omega}{\exp(\hbar\omega/kT) - 1} \int_0^\infty \frac{2\pi^2 m_g \mu}{\kappa^2 \hbar^2} \frac{\sinh 2\pi p_i \sinh 2\pi p_f}{(\cosh 2\pi p_i - \cosh 2\pi p_f)^2}$$

$$\times \frac{(A_i + A_f)^2}{A_i A_f} \exp \frac{-E_i}{kT} dE_i \tag{3.16}$$

where μ is the mass ratio m_g/m_s. This expression corresponds, apart from
slight differences in notation, to that derived by LJD [34], and is discussed
in the following section.

3.1.2. Validity of the LJD Theory, and Recent Developments

We have followed through the LJD calculation of the accommodation coefficient because of its historical significance and because it has provided the
basis for several other studies. We now consider the question of the validity of some of the assumptions used in the above analysis, and discuss some
of the more recent developments.

The first thing to note about the LJD analysis is that it is a perturbation
treatment, based on a perturbation about a condition which corresponds to
purely specular scattering. Thus we should only expect it to be valid when
the overall probability of the gas atom being scattered out of the specular
state is small. In this connection we note that the perturbation treatment
does not conserve the number of gas atoms, and if the scattering probability out of the specular state is not small then the final number of gas atoms
may be considerably greater than the initial number. A modification of the
above treatment has been put forward by Goodman [27], providing a unitary
theory for the scattering, i.e., one in which the total scattering probability

is unity. The corrected unitary scattering probability for a particular transition is obtained by dividing the nonunitary probability by the factor $(1 + \frac{1}{4} P_T)^2$, where P_T is the total (nonunitary) probability for scattering out of the initial state into a nonspecular final state. On this basis Goodman suggests that the first-order treatment should be reasonably valid when $P_T < 0.2$.

Another assumption which has been made is that in the expansion of the perturbing Hamiltonian only the linear term (first derivative of the potential) has been kept, as in Eq. (3.3). As a direct result of this restriction only transitions involving a single-phonon change in any mode are allowed. The question of whether it is valid to sum (or integrate) over all modes to get the total energy transfer as in the LJD analysis has been a matter of some debate, but from Gilbey [44] we see that, within the limitations of the first-order perturbation treatment (i.e., the transition probability must be small), this procedure is valid. The effect of keeping the second-order term in the expansion of the potential [Eq. (3.3)], although still using first-order perturbation theory, has been investigated by Allen and Feuer [45]. This modification allows the possibility of two-phonon transitions to a single mode. Allen and Feuer have numerically evaluated the ratio $\alpha(2)/\alpha(1)$, where $\alpha(2)$ is the contribution of two-phonon transitions to the accommodation coefficient, and $\alpha(1)$ is the single-phonon contribution. For helium on various metals at $300°K$, they find values for this ratio ranging from 0.005 to 0.09. It has been pointed out by Goodman [27], however, that this calculation is only valid if the calculation of $\alpha(1)$ itself is valid. Goodman shows that if one makes the unitarity correction for $\alpha(1)$, then $\alpha(1)$ is significantly reduced and the ratio $\alpha(2)/\alpha(1)$ attains values as high as 3.4 (for He/K). Two-phonon processes may therefore be very important for some systems around room temperature, and even more so at higher temperatures.

This leads up to the question of the classical limit. At higher energies, when many phonons are exchanged at each collision, one would expect the collision process to be rigorously described by classical mechanics. In this high energy limit (impulsive collisions) the accommodation coefficient is given by the well known expression due to Baule [46]:

$$\alpha = 4\mu/(1 + \mu)^2$$
$$\simeq 4\mu \qquad\qquad (3.17)$$

for $\mu \ll 1$. One would hardly expect a perturbation analysis which allows
only a single-phonon transition to each mode to yield this classical limit. In
fact, however, in the limit $T \to \infty$, Eq. (3.16) gives $\alpha_{lim} = 4\mu$ in agreement
with Eq. (3.17) for $\mu \ll 1$ [44]. One aspect of this result can be understood
by considering the classical analog of the perturbation treatment, in which
one derives a zero-order trajectory (of the gas atom) and corresponding force
pulse with the lattice assumed fixed, and then applies this force pulse to the
lattice to obtain the energy transfer. The high energy limit in this case corresponds to zero collision time, and in this case the procedure becomes exact. There remains, however, the restriction to single-phonon transitions,
which should have prevented the LJD expression giving the classical limit.
This point has been discussed by Gilbey [44]; in the formalism the restriction to single-phonon transitions is overcome by the occurrence of greater-
than-unity probabilities for these transitions. Thus the fact that the LJD expression gives the correct classical limit must be regarded as fortuitous,
and is another example of the well known but not well understood fact that
the average energy transfer to a harmonic oscillator is given to be the same
by both classical and quantum mechanics, however large the perturbation
[47].

   In addition to the assumptions associated with the perturbation treatment,
there is another major assumption involved in the LJD theory, i.e., the one-
dimensional (1D) assumption. We considered only the component of the motion of the gas atom normal to the surface, and assumed a head-on collision
with a single surface atom. The only concession to the true 3D nature of
the process is the assignment of a Debye frequency spectrum to the normal
modes. This is a convenient point at which to introduce the matter of the
agreement between the LJD theory and experiment. It is probably true to
say that no really satisfactory agreement between experiment and the LJD
theory has been obtained. This has been discussed by Gilbey [44], who
suggested that the shortcoming of the LJD theory is related to the basically
different response characteristics of 1D and 3D lattices. This point has
been taken up in depth by Goodman [27] and by Goodman and Gillerlain
[48]. It is found that the LJD theory tends to overestimate the transition
probabilities. The likely reason for this, suggested in Ref. [27], is that
the 1D LJD theory does not take into account the law of conservation of

tangential (i.e., parallel to the surface) momentum of the gas atom plus phonons, which would apply in a true 3D treatment. As a result of neglecting
this conservation law many transitions are allowed in the 1D model which
would not be allowed in the 3D case, and hence the overall transition probabilities are overestimated.

We also note that Eq. (3.16) only allows for transitions between initial
and final free states of the gas atom. In many cases the trapping of the gas
atom at the surface (i.e., transition from a free to a bound state) is quite
important, although one probably cannot hope to treat such cases accurately
with a first-order perturbation approach. A new 3D version of the LJD
theory, which takes into account bound-state transitions, is under development [48]. A unitary one-phonon treatment of inelastic scattering has been
presented by Manson and Celli [49], which is based on a treatment presented
in connection with elastic scattering in the next section.

3.1.3. Elastic Scattering Theory

To study elastic scattering the simplest model to use is a rigid lattice
with the surface atoms fixed at their equilibrium positions. The gas atom
then moves in a fixed potential field. This type of model has been studied in
detail by Lennard-Jones and Devonshire (LJD) [31-33]; they considered a
simple square lattice of surface atoms with a Morse potential as in Eq. (3.1)
acting between the gas atom and each surface atom. The important features
which come out of this type of analysis are the position and intensity of the
diffraction peaks, and the phenomenon of selective adsorption, which shows
up as anomalous dips in the diffraction peaks.

The potential field is periodic in the (x, y) directions parallel to the surface, and has the general form

$$V(x, y, z) = \sum_{m, n} B_{mn}(z) \exp \left\{ 2\pi i [(mx/a) + (ny/a)] \right\} \qquad (3.18)$$

where m and n are integers and a is the lattice spacing. The asymptotic
form (z → ∞) of the wavefunction corresponding to this potential follows

from the Schrödinger equation and the condition of conservation of energy, as given in general form by Beder [6]:

$$\psi = \exp\,(i\underline{k_i} \cdot \underline{r}) + \sum_{m,n} A_{mn} \exp\,i \left\{ \left[k_{xi} + \frac{2\pi\,m}{a} \right] x + \left[k_{yi} + \frac{2\pi\,n}{a} \right] y + k_{zf} z \right\} \quad (3.19)$$

where $(k_{xi},\ k_{yi},\ k_{zi})$ are the (x, y, z) components of $\underline{k_i}$. This expression gives the conditions for the diffraction peaks

$$k_{xf} = k_{xi} + \frac{2\pi\,m}{a}$$

$$k_{yf} = k_{yi} + \frac{2\pi\,n}{a} \quad (3.20)$$

where k is related to the wavelength through $k = 2\pi/\lambda$. These diffraction conditions can also be written in terms of the incident and outgoing angles as shown in Fig. 3.2,

$$\cos\,\alpha_r - \cos\,\alpha_i = \frac{m\lambda}{a}$$

$$\cos\,\beta_r - \cos\,\beta_i = \frac{n\lambda}{a} \quad (3.21)$$

The diffraction peaks are designated by (m, n). The specular $(0, 0)$ and first-order $(0, \pm 1)$ peaks were observed in the well known experiments of Estermann and Stern [50] which confirmed the wave nature of atoms. These experiments and later ones are reviewed by Massey and Burhop [51]. In these later experiments Frisch and Stern [52] observed the phenomenon of selective adsorption, and the theoretical interpretation of these results by Lennard-Jones and Devonshire [31-33] represents one of the most successful theoretical studies in the field of gas-surface scattering.

The LJD analysis of selective adsorption is based on a first-order perturbation treatment of the elastic scattering. The interaction potential, Eq. (3.18), is used in the simplified form

$$V(x, y, z) = V(z) + 2U_1(z) \left(\cos\,\frac{2\pi\,x}{a} + \cos\,\frac{2\pi\,y}{a} \right) \quad (3.22)$$

The periodic term is treated as a perturbation:

$$H' = 2\,U_1(z) \left(\cos\,\frac{2\pi\,x}{a} + \cos\,\frac{2\pi\,y}{a} \right) \quad (3.23)$$

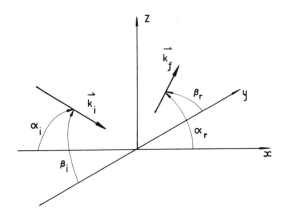

FIG. 3.2. Geometry to illustrate the diffraction condition, Eq. (3.21).

The potential $V(z)$ allows the possibility of bound states (corresponding to $k_{zf}^2 < 0$). Consideration of the transition matrix $<\psi_f(\text{bound})|H'|\psi_i(\text{free})>$ for transition from free states to bound states leads to the following diffraction-type selection rules.

$$k_{yf} = k_{yi} \quad \text{and} \quad k_{xf} - k_{xi} = \pm 2\pi/a \qquad (3.24)$$

or

$$k_{xf} = k_{xi} \quad \text{and} \quad k_{yf} - k_{yi} = \pm 2\pi/a \qquad (3.25)$$

These rules combined with the condition of energy conservation allowing for the bound state

$$k_{xi}^2 + k_{yi}^2 + k_{zi}^2 = k_{xf}^2 + k_{yf}^2 - |k_{zf}^2| \qquad (3.26)$$

lead to relations of the form

$$k_{zi}^2 = \pm \frac{4\pi}{a} k_{yi} + \left(\frac{2\pi}{a}\right)^2 - \frac{2m_g}{\hbar^2}|E_n| \qquad (3.27)$$

where E_n is the bound state energy eigenvalue. This parabolic relationship between k_{yi} and k_{zi} (for transition into bound states) was confirmed in the experimental observations of Frisch and Stern [52]. By matching with these experimental results LJD obtained values for E_1 and E_2, and hence were able to deduce values for the parameters κ and D in the Morse potential (shown in Table 3.1).

TABLE 3.1

Values of Energy Levels E_n and Morse Potential Parameters κ and D

Obtained by LJD by Comparison with Selective Adsorption Results

System	$E_n/k,$ $^\circ K$	$\kappa,$ nm^{-1}	$D/k,$ $^\circ K$
He/LiF	-25	11.0	87
	-64		
He/NaF	-40	14.4	133
	-96		

The first-order LJD theory also yields the intensities of the free-free diffracted beams. The first-order correction to the eigenvector $\psi^{(1)}$ contains four terms corresponding to the diffraction peaks $(\pm 1, 0)$ and $(0, \pm 1)$. The intensities in the various beams are given by expressions of the form

$$R_{mn} = \frac{k_{zf}}{k_{zi}} |A_{mn}|^2 \qquad (3.28)$$

where

$$A_{mn} = \exp (ik_{zf}z) \frac{1}{2k_{zf}} <k_{zf}| U_1(z) |k_{zi}> \qquad (3.29)$$

and $U_1(z)$ is defined by Eq. (3.22).

The results so far described in this Section were all obtained in the 1930's. The next relevant study was that of Beder [53]; he obtained the exact scattering distribution corresponding to a 2D model with a square-well approximation to the gas-surface interaction potential. Tsuchida [54] has given a theoretical treatment of elastic scattering, more general than the original LJD analysis, but still basically a perturbation treatment. He considers specifically the system He/LiF, and makes numerical calculations for this case.

The most comprehensive of the recent treatments is that of Cabrera, Celli, Goodman, and Manson [55] (hereafter referred to as CCGM) and Goodman [56], and we shall concentrate attention on these. These authors present a treatment based on exact quantum mechanical scattering theory, which incorporates many phonon processes, and can be extended to cover inelastic scattering [49]. The theory developed in Ref. [55] is quite lengthy, so we here consider just the main features, and the important results. The method involves splitting the interaction potential into two parts, a "large" part giving total specular reflection, and a remainder which is handled formally exactly in terms of a reduced transition matrix [57, 58].

The surface atoms are assumed to form a perfect, two-dimensional, infinite periodic array. These assumptions result in a two-dimensional reciprocal lattice, each vector of which is parallel to the surface; the reciprocal lattice vectors are denoted by $\underline{G}, \underline{G}'$. CCGM then define

$$v_{\underline{G}}(z) = L^2 \int_{surface} v(\underline{r}) \; \exp \, (-i\underline{G} \cdot \underline{R}_t) \; d\underline{R}_t \qquad (3.30)$$

where L^2 is the surface area, $v(\underline{r})$ is the gas-surface interaction potential averaged over the thermal motion of the surface atoms, and $\underline{R}_t$ is the two-dimensional position vector (x, y) parallel to the surface. The wave function of the gas atom can be expanded:

$$\psi(\underline{r}) = \sum_{\underline{G}} \psi_{\underline{G}}(z) \, \exp \, [(i\underline{K}_i + \underline{G}) \cdot \underline{R}_t] \qquad (3.31)$$

where $\underline{K}_i$ is the component of the gas atom wave vector $\underline{k}_i$ parallel to the surface. Using Eqs. (3.30) and (3.31) in the gas atom Schrödinger equation yields the relation.

$$\left(\frac{d^2}{dz^2} + k_{\underline{G}z}^2 - \frac{2m_g}{\hbar^2} v_{\underline{0}} \right) \psi_{\underline{G}} = \frac{2m_g}{\hbar^2} \sum_{\underline{G}' \neq \underline{G}} v_{\underline{G} - \underline{G}'} \, \psi_{\underline{G}'} \qquad (3.32)$$

where

$$k_{\underline{G}z}^2 = k_i^2 - (\underline{K}_i + \underline{G})^2 \qquad (3.33)$$

and $v_0(z)$ is $v(\underline{r})$ averaged over the directions x and y parallel to the surface. The potential $v_0(z)$ has associated with it a complete set $\phi_\alpha(z)$ of eigenstates, including both bound (negative energy) and continuum (positive energy) states, given by the equation

$$\left(\frac{d^2}{dz^2} + \alpha^2 - \frac{2m_g}{\hbar^2} v_0(z)\right) \phi_\alpha(z) = 0 \tag{3.34}$$

where α is the appropriate wave vector [55].

Bearing in mind that the object of the exercise is to find $\psi_{\underline{G}}$ and hence the intensities of the various diffracted beams, $\psi_{\underline{G}}$ is expanded in terms of the ϕ_α:

$$\psi_{\underline{G}} = \sum_\alpha C_{\underline{G}\alpha} \phi_\alpha \tag{3.35}$$

Inserting Eq. (3.35) into Eq. (3.32) yields a relation between the $C_{\underline{G}\alpha}$. Introducing a t matrix defined by

$$(\underline{G}\alpha|t|\underline{0}k_{zi}) \exp i\xi_0 = \sum_{\underline{G}' \neq \underline{G}} \sum_\beta C_{\underline{G}'\beta} (\alpha|v_{\underline{G}-\underline{G}'}|\beta) \tag{3.36}$$

where ξ_0 is a phase shift related to the specular scattering, CCGM then obtain an expression for $C_{\underline{G}\alpha}$ which, when substituted back into Eq. (3.35), yields

$$\psi_{\underline{G}} \exp(-i\xi_0) = \phi_{k_{0z}} \delta_{\underline{G},\underline{0}} + \frac{2m_g}{\hbar^2} \sum_\alpha \frac{(\underline{G}\alpha|t|\underline{0}k_{zi})\phi_\alpha}{(k_{\underline{G}z}^2 - \alpha^2 + i\epsilon)} \tag{3.37}$$

In the limit $z \to \infty$ CCGM are able to carry out the summation over α in Eq. (3.37), and hence obtain the following expressions for the wave functions:

$$L^{1/2}\psi_{\underline{0}} (z \to \infty) = \exp(-ik_{zi}z) + \left[1 - \frac{im_g L}{\hbar^2 k_{zi}} (\underline{0}k_{zi}|t|\underline{0}k_{zi})\right]$$

$$\cdot \exp\left[i(k_{zi}z + 2\xi_0)\right] \tag{3.38}$$

and

$$L^{1/2}\psi_{\underline{F}} (z \to \infty) = -\frac{im_g L}{\hbar^2 k_{Fz}} \left(\underline{F}k_{Fz}|t|\underline{0}k_{zi}\right) \exp\left[i(k_{\underline{F}z}z + \xi_0 + \xi_{\underline{F}})\right] \tag{3.39}$$

where $\underline{F}$ denotes a value of $\underline{G}$ for which a diffracted state occurs.

To proceed further one needs to know the t matrices appearing in Eqs. (3.38) and (3.39). Inserting the expressions for $C_{\underline{G}'\beta}$ into Eq. (3.36) yields an equation for the t matrix. Up to this point the analysis has been exact;

to solve this equation for the t matrix CCGM introduce a mathematical approximation (neglecting certain poles in the integration over continuum states), the physical significance of which is not fully apparent. The t matrix can then be found in terms of matrix elements of the potential components $v_{\underline{G}}$. Using Eqs. (3.38) and (3.39) CCGM calculate for various special cases the intensity of the outgoing beams

$$R_{\underline{F}} \equiv L \, (k_{\underline{F}z}/k_{zi}) \, |\psi_{\underline{F}}^{+} \, (z \to \infty)|^{2} \tag{3.40}$$

where $\psi_{\underline{F}}^{+}$ denotes the outgoing part of $\psi_{\underline{F}}$.

We note that the CCGM theory satisfies the conservation of atoms condition

$$\sum_{\underline{F}} R_{\underline{F}} = 1 \tag{3.41}$$

which follows from the unitarity condition of the t matrix theory. As discussed in Section 3.1.2 the usual first-order perturbation treatments do not satisfy this condition.

Two interesting general features emerge from the CCGM analysis. The first feature is related to the phenomenon of selective adsorption discussed above. CCGM show that in this case of "resonance" of the incident beam with a bound state the intensity of the specular beam rises sharply, while that of each of the other beams falls. In the experimental observations [52], however, the intensity of the specular beam falls as resonance is approached. CCGM show that this fall in specular intensity is almost certainly due to inelastic scattering of the gas beam (which is more probable when the atoms move over the surface for some time in a bound state), as was originally suggested by Lennard-Jones and Devonshire [31-33]. The second feature is called by CCGM "surface resonance"; this refers to the behavior of a diffracted beam as it just appears above, or just disappears below, the surface. The behavior depends on the value of the parameter $d \equiv (2m_g D)^{1/2}/\hbar\kappa$; as mentioned in Section 3.1.1 the integer nearest to d gives the number of bound states in the Morse potential. When d equals half an odd integer R_0 jumps discontinuously to unity and all the $R_{\underline{F}}$ ($\underline{F} \neq 0$) fall to zero as $k_{\underline{F}z}^{2}$

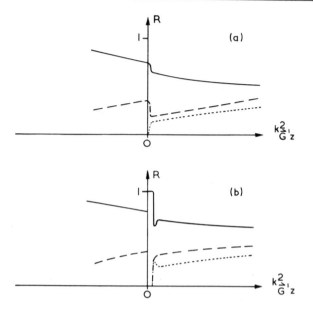

FIG. 3.3. Behavior of diffracted beam intensities R_0(——), R_F(----) and $R_{F'}$($\cdots$) as functions of $k_{\underline{G}'z}^2$. (a) R_0, $R_{F'}(\underline{F}' = \underline{G}')$ and R_F $(\underline{F} \neq \underline{G}')$ as the diffracted beam $\underline{F}'$ appears above the surface (at $k_{\underline{G}'z}^2 = k_{\underline{F}'z}^2 = 0$) when $d \simeq$ integer. (b) As for (a) when d = half an odd integer (Cabrera et al. [55]).

changes from a negative value to zero. As $k_{\underline{F}z}^2$ increases, R_0 decreases to a minimum, while $R_{\underline{F}'}$ shows a corresponding maximum. The overall behavior is shown in Fig. 3.3(b), reproduced from Ref. [55]. The more normal behavior, for d not equal to half an odd integer, is shown in Fig. 3.3(a) (for $d \simeq$ integer). In this case as the diffracted beam F' disappears the intensity $R_{\underline{F}'}$ falls rapidly but smoothly to zero while all the other intensities $R_{\underline{F}}$ ($F \neq F'$) increase smoothly onto new curves. For values of d between the two extreme cases considered there is a continuously varying range of intermediate behavior.

At the same time as these theoretical developments some important new diffraction experiments have been carried out [16, 17, 59, 60], taking advantage of modern techniques to give much more detailed results than were available previously. These experiments mainly involve the gases ^{3}He, ^{4}He, H_2, and D_2 scattered from LiF, and have yielded accurate results for the location and intensity of the diffracted beams as a function of beam temperature and angle of incidence. The qualitative predictions of the CCGM theory have been shown by Goodman [56] to be in reasonable agreement with the experimental results for ^{3}He and ^{4}He/LiF.

3.2. CLASSICAL LATTICE THEORIES

During the past ten years a great deal of work has been done [61-77] on models of the gas-surface collision in which the solid is represented by a lattice, either a one-dimensional chain or a three-dimensional cubic lattice, and classical mechanics is used for the analysis. Although great progress has been made in the mathematical analysis of these models and good agreement with experiment has apparently been obtained, the situation with regard to these models is not altogether satisfactory and the emphasis has tended to shift recently to other models. The trouble is that it has proved very difficult to take surface temperature effects into account in analyzing the interaction of the gas atom with the lattice. As a result the experimental results have been compared with theoretical results in which the solid is initially at $0°$K and, as discussed in detail in Section 3.2.2, this is a rather dubious procedure. Before considering the general question of the validity of the lattice models, and comparison with experiment, we give in Section 3.2.1 a brief outline of the mathematical analysis of these models.

3.2.1. Mathematical Analysis of Lattice Models

The earliest lattice models considered a simple one-dimensional chain. More recent treatments have used two-, three- and also general n-dimensional lattices. In this section we concentrate on a 3D lattice, which is obviously the most physically realistic case and is most likely to produce useful results.

Initially we consider an infinite 3D lattice, and later obtain the result for the more physically realistic semi-infinite lattice. For a lattice spacing a the equilibrium coordinate of the general atom is

$$x_j = a\, N_j \tag{3.42}$$

$(N_j = -\infty, \ldots, -1, 0, 1, \ldots, \infty$ for $j = 1, 2, 3)$. This is conveniently written in a vector notation

$$\underline{x} = a\underline{N} \tag{3.43}$$

in which both sides are 3D column vectors. We consider a lattice with central and noncentral forces acting between nearest neighbors only. Thus a nearest neighbor in the jth lattice direction exerts a force in the j direction with central force constant k_{jj}, and a nearest neighbor in the ith direction ($i \neq j$) exerts a force in the jth direction with noncentral force constant k_{ji}. Since the present object is to demonstrate the method of analysis we make the simplification $k_{ji} = k_0$ for all i and j. The case without this restriction can be tackled as in Refs. [69] and [72].

We are using classical mechanics throughout, so we begin by setting down the equations of motion of the lattice. For the case where a force F(t) acts in the x_3 direction on the atom at the origin the equation of motion of the atom with equilibrium position $a\underline{N}$ in the x_3 direction can be written

$$m_s \, \ddot{x}_3 \, (\underline{N}, t) = \sum_{j=1}^{3} k_0 [x_3(\underline{N} - \hat{\underline{x}}_j, t) + x_3(\underline{N} + \hat{\underline{x}}_j, t)] - 6k_0 x_3 \, (\underline{N}, t)$$

$$+ F(t)\, \delta_{\underline{0}\,\underline{N}} \tag{3.44}$$

where $\hat{\underline{x}}_j$ is a unit vector in the j direction.

We are interested in the case where the x_3 direction, in which the force F(t) is applied, is perpendicular to the surface, and to be consistent with the notation in the remainder of this article we replace x_3 by z and the unit vector $\hat{\underline{x}}_3$ by $\hat{\underline{z}}$.

We also introduce a dimensionless time defined by

$$\tau \equiv 2 \, (k_0/m_s)^{1/2} \, t \tag{3.45}$$

By using the normal mode transform

$$G(\underline{\theta}, \tau) = \sum_{\underline{N}} \exp \, (i\underline{N} \cdot \underline{\theta}) \, z(\underline{N}, \tau) \tag{3.46}$$

the equation of motion Eq. (3.44) reduces to an ordinary differential equation in G. By taking the Laplace transform of this equation for G, one obtains the following expression [65] for the displacement z:

$$z(\underline{N}, \tau) = \frac{1}{4} k_0^{-1} \int_0^\tau F(\tau - t) \, \chi(\underline{N}, t) \, dt \tag{3.47}$$

where $\chi(\underline{N}, t)$ is the response of the infinite 3D lattice atom $\underline{N}$ to unit initial velocity of the atom $\underline{0}$. This response function is given by [65]

$$\chi(\underline{N}, \tau) = (1/2\pi)^3 \int_0^{2\pi} \ldots \int_0^{2\pi} \cos(\underline{N}, \underline{\theta}) \sin[\tau\omega(\underline{\theta})] \, \omega^{-1}(\underline{\theta}) \, d\underline{\theta} \tag{3.48}$$

The analysis so far has been based on a 3D infinite lattice. We are mainly interested in the response $X(\tau)$ of a surface atom of a 3D semi–infinite lattice. As shown by Goodman [65] this response function is related to the response function in Eq. (3.48) as follows

$$X(\tau) = \chi(\underline{0}, \tau) + \chi(\underline{0} + \underline{z}, \tau) \tag{3.49}$$

Thus the response $z_S(\tau)$ of the surface atom to the force $F(\tau)$ applied to this atom is given by

$$z_s(\tau) = \frac{1}{4} k_0^{-1} \int_0^\tau F(\tau - t) \, X(t) \, dt \tag{3.50}$$

The quantities which are important for calculations of the energy accommodation coefficient are $X(\tau)$, $\dot{X}(\tau)$ and $\int_0^\tau X(t) dt$, and we now consider the calculation of these quantities. From Eqs. (3.48) and (3.49) we obtain

$$\dot{X}(\tau) = 2(1/2\pi)^3 \int_0^{2\pi} \ldots \int_0^{2\pi} \cos^2\left(\frac{1}{2}\theta_1\right) \cos[\tau\omega(\underline{\theta})] \, d\underline{\theta} \tag{3.51}$$

Expanding $\cos \tau\omega$ as a power series in $\tau\omega$ and using the frequency spectrum

$$\omega^2(\underline{\theta}) = \sum_{j=1}^3 \sin^2\left(\frac{1}{2}\theta_j\right) \tag{3.52}$$

Eq. (3.51) becomes

$$\dot{X}(\tau) = \frac{2}{\pi^3} \sum_{m=0}^\infty (-1)^m \frac{\tau^{2m}}{(2m)!!} \int_0^\pi \int_0^\pi \int_0^\pi \cos^2 \zeta_1$$

$$\cdot (\sin^2 \zeta_1 + \sin^2 \zeta_2 + \sin^2 \zeta_3)^m \, d\zeta_1 d\zeta_2 d\zeta_3 \tag{3.53}$$

where $\underline{c} = \frac{1}{2}\underline{\theta}$. Integrating Eq. (3.53)

$$\dot{X}(\tau) = \sum_{m=0}^{\infty} \beta_m \tau^{2m} \tag{3.54}$$

where

$$\beta_m = (-1)^m \frac{m!}{(2m)!\,4^m} \sum_{k=0}^{m} \sum_{\ell=0}^{k} \frac{(2m-2k)!\,(2k-2\ell)!\,(2\ell)!}{(m-k)!^2\,(m-k+1)!\,(k-\ell)!^3\,\ell!^3} \tag{3.55}$$

The quantities $X(\tau)$ and $\int_0^{\tau} X(t)dt$ can then be evaluated by term-by-term integration of Eq. (2.54). This series solution is only useful for $\tau \lesssim 8$. For large $\tau\,(\tau \gtrsim 15)$ the following asymptotic expression is valid [65]:

$$X(\tau) \simeq \left(\frac{\pi^3 \tau^3}{32}\right)^{-1/2} \left[2 \sin\left(\tau + \frac{\pi}{4}\right) + 2^{1/4} \sin\left(\sqrt{2}\tau - \frac{\pi}{4}\right)\right] \tag{3.56}$$

The behavior of $X(\tau)$ for $0 \leq \tau \leq 20$ is shown in Fig. 3.4, reproduced from Ref. [65]. In the same figure the corresponding response function for the 1D lattice is also shown, and it is seen to be qualitatively quite different. This difference in behavior of 1D and 3D lattices is discussed in more detail in Section 3.2.2.

The preceding derivation of the lattice response function used the normal mode transform given in Eq. (3.46). An alternative mathematical treatment which uses a Laurent-Cauchy transform has been given by Chambers and Kinzer [74]. This alternative method offers some advantages, particularly as far as the boundary conditions are concerned, and one does not need to use a relation as in Eq. (3.49) to obtain results for a semi-infinite lattice. Since the end result of the two methods is the same, and the normal mode treatment has been more widely used, we have concentrated here on that method.

The next step is to describe the collision of a gas atom with the solid surface in terms of the response function derived above. We consider only the component of the motion of the gas atom normal to the surface, and assume it has a line of centers collision with the atom $\underline{0}$ (see Fig. 3.1). The

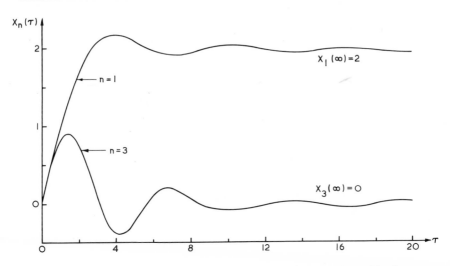

FIG. 3.4.   Position response functions for 1D and 3D lattices (Goodman [65]).

lattice is assumed to be initially stationary, and the displacement of the atom $\underline{0}$ from its equilibrium position is denoted by z_s.   The position of the gas atom m_g with respect to the equilibrium position of this surface atom is z_g; we further define a relative coordinate

$$r = z_s - z_g \tag{3.57}$$

The interaction between the gas atom and the lattice atom $\underline{0}$ is given by $V(r)$, and hence the force on the lattice atom due to the gas atom is

$$F(\tau) = - \partial V/\partial r \tag{3.58}$$

The displacement of the lattice atom is related to this force by Eq. (3.50) derived above.   Using Eq. (3.50) and its time derivatives in the equation of motion of the gas atom

$$\ddot{z}_g = - F(t)/m_g \tag{3.59}$$

the required equation of motion in the relative coordinate is obtained:

$$r(\tau) = \frac{1+\mu}{4\mu} \ F(\tau) + \frac{1}{4} \int_0^\tau \ddot{X}(t) \ F(\tau - t) \ dt \tag{3.60}$$

This is the basic equation describing the gas-surface collision. It can only be solved analytically in a few special cases. Zwanzig [62] has given two cases in which the corresponding equation for the 1D lattice can be solved exactly. These cases are for two particular choices of the gas-surface potential: the truncated harmonic and the "constant force" potential. Since neither these potentials nor the 1D lattice are physically realistic we do not consider these exact solutions further. For a 3D lattice, and with a truncated harmonic potential, Chambers and Kinzer [74] give an analytic solution for $r(\tau)$ in Eq. (3.60) in terms of a Neumann series, by expanding both $r(\tau)$ and the $X(t)$ in terms of Bessel functions. Again, however, because this solution is restricted to a potential which is not physically realistic we do not consider further details of it here. The most comprehensive results using physically realistic potentials have been obtained using numerical methods to solve Eq. (3.60).

Two different procedures have been used for these numerical solutions. The first procedure uses the perturbation method which we have already come across in Sections 2 and 3.1. In this method the gas atom trajectory is calculated with the lattice assumed fixed, and hence a zero-order expression for the force $F_0(t)$ is found. This expression is substituted into the right-hand side of Eq. (3.60), and the integral carried out numerically to find $r(\tau)$. The $r(\tau)$ thus found should be a good approximation to the exact solution when $\mu \ll 1$. More explicitly the energy transferred to the lattice (i.e., lost by the gas atom) is given by [using Eq. (3.50)]

$$\Delta E = \int_{-\infty}^{\infty} \int_{0}^{\infty} F_0(\tau)\, F_0(\tau - \tau')\, X(\tau')\, d\tau\, d\tau' \qquad (3.61)$$

The second procedure which has been used is valid for any value of μ, and any magnitude of energy transfer (including trapping) and involves a step-by-step integration of the equation of motion of the gas atom [Eq. (3.59)], together with the response of the surface atom [Eq. (3.50)] and the force as a function of the relative coordinate r. For details of this step-by-step integration, for the case of a Morse potential for the gas-surface interaction, see Ref. [68].

3.2.2. Validity of Lattice Models and Comparison with Experiment

As has been indicated in Section 3.2.1 the classical lattice models have been used almost exclusively for the calculation of energy accommodation coefficients (and to some extent trapping probabilities; we return to this in Section 4), rather than the molecular beam scattering results as were described in Section 2. The main reasons for this are twofold: first historical, because at the time of the major work on these theories the most reliable data were those on the accommodation coefficient; and second because the scattering data obviously show effects dependent on the surface temperature, and this has not been incorporated in the lattice theories.

On the question of the validity of the lattice models the first point to consider is the question of 1D versus 3D models. As shown in Fig. 3.4 the response function for the 1D lattice is quite different from that for the 3D lattice. The difference between the two response functions is most marked for collision times larger than the inverse natural frequency of the solid ($\tau > 1$). As shown in Ref. [65] the equilibrium displacement $z_s(t)$ of a surface atom of the 3D lattice under the action of a slowly varying force $F(t)$ is given by

$$z_s(t) \simeq 0.34 \, \frac{F(t)}{k_0} \tag{3.62}$$

and hence for a sufficiently slowly varying force the velocity of the surface atom tends to zero, $\dot{z}_s(t) \simeq 0$. The 1D lattice has no such equilibrium position, and in fact gives way indefinitely under the action of a constant force F_0, the constant velocity of the surface atom being given by

$$\dot{z}_s(t) \simeq \frac{F_0}{(k_0 m_s)^{1/2}} \tag{3.63}$$

It follows that a slowly varying force is very efficient at transferring energy to a 1D lattice, but transfers very little energy to a 3D lattice. This obviously has important implications for the calculation of the energy accommodation coefficient. The 1D model will not give correct results, except for the high-speed limit for small μ, when only the surface atom is involved

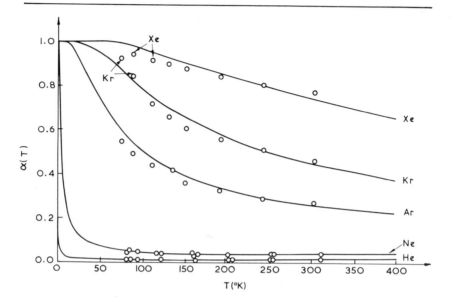

FIG. 3.5. Comparison between results from lattice theory and experimental measurements of the energy accommodation coefficient (Goodman
[68]).

in the collision and the collision time is short. These particular restrictions
on the 1D model do not apply to the hard-cube and soft-cube models described
in Section 3.4.

From now on we will concentrate attention on the 3D lattice models. An
example of comparison between theory and experiment is shown in Fig. 3.5,
reproduced from Ref. [68]. The experimental results in the figure (circles)
are from Thomas and Schofield [78] and Thomas [79] for the gases He, Ne,
Ar, Kr, and Xe on W. These experimental results are obtained by measuring
the heat transfer from a tungsten filament in a low pressure system. As discussed in Section 2, the filament is kept slightly hotter than the background
gas temperature, so that the net heat transfer is from the solid to the gas.
The lines passing through the experimental points in Fig. 3.5 are given by
the 3D lattice model of Goodman, as was outlined in Section 3.2.1. The parameters of the Morse potential κ and D which enter into the theory have been

chosen to provide best fit with the experimental results. The κ and D values so obtained are reasonable in the sense that the inverse ranges κ are similar to those found in gas-gas interactions, and the energies D compare well with the values of the heat of adsorption (where these are available). It is seen from Fig. 3.5 that the agreement between theory and experiment is excellent.

There is, however, one major inconsistency in this comparison; referring to the definition of α in Eq. (2.2), both the quantities $\bar{E}_r - \bar{E}_i$ and $\bar{E}_s - \bar{E}_i$ are positive quantities (α is of course also a positive quantity) in the experimental case. In the theoretical case, on the other hand, the surface is assumed to be initially at absolute zero and hence both the quantities $\bar{E}_r - \bar{E}_i$ and $\bar{E}_s - \bar{E}_i$ are negative. In other words in the experimental case the net energy transfer is from the solid to the gas, whereas in the theoretical case it is from the gas to the solid. The comparison between theory and experiment is therefore basically invalid. This comparison procedure could, however, be justified if α is independent of surface temperature for a given gas temperature. There is some evidence (as discussed, for example, in Refs. [80] and [81]) that this might be the case, at least at higher temperatures, but unless it can be shown to be generally true down to $0°K$ the good agreement with experiment is probably fortuitous and the usefulness of the lattice theories is severely limited.

Since the above-mentioned drawbacks of the lattice models stem from omission of surface temperature effects, we now consider briefly the attempts which have been made to include such effects. The inclusion of the thermal motion of the lattice is a statistical problem. This has been tackled formally by Goodman [69] (for the 3D lattice) and by Armand [76] (for the 1D lattice). Using the condition of equipartition of energy between the modes of the lattice at thermal equilibrium, probability distributions are assigned to the normal mode coordinates and velocities, and to the corresponding phases. The formal methods for using these distributions to calculate quantities such as the energy accommodation coefficient are quite straightforward and are described in Refs. [69] and [76]. In neither of these cases have calculations been made of α or the scattering distributions to enable comparison with experiment; the main limitation is the large amount of numerical computation required.

We should also mention the empirical formula for the energy accommodation coefficient given by Goodman and Wachman [77]. This formula is
based on the qualitative behavior of α given by Goodman's lattice theory
[68] (as described above), and also takes other effects into account, including nonnormal and non-head-on collisions. The formula contains several free parameters with can be chosen for a given gas-solid system to give
good agreement with a wide range of experimental data.

3.3. CLASSICAL CONTINUUM THEORIES

One of the first comprehensive studies of the gas-surface collision problem was given by Landau [82], using an elastic continuum representation of
the solid. Since this early study the continuum model has received relatively little attention compared with, for example, the classical lattice
theories (Section 3.2). There are now signs [81] of renewed interest in
the continuum model, as is discussed below.

Before considering specific theoretical treatments we consider the likely
range of validity, and the practical advantages of the continuum analysis.
While we cannot lay down precise conditions for the validity of the continuum
theory, we certainly expect the representation of the solid by an elastic continuum to be reasonably valid for collisions in which the interaction time is
long compared with the Debye oscillation period of the solid. Another way
of expressing this condition is to say that the distance that the disturbance
caused by the impact propagates into the solid during the collision time
should be much greater than a lattice spacing. In either form the condition can be expressed by $(b/u_{ni}) \gg \nu_D^{-1}$, where b is the range parameter of
the repulsive potential, u_{ni} is the normal component of the incident gas velocity, and ν_D is the Debye frequency of the solid; the left-hand side of this
expression is a measure of the collision time. In fact for gas-surface collisions in the thermal energy range, as we are mainly concerned with here,
it is found that $b\nu_D/u_{ni} \simeq 1$. Thus we do not expect the continuum model to
be valid in many cases of interest, although it may still yield useful results.

Various advantages of the continuum model have been suggested by Trilling [81], some of which are as follows:

(1) The model represents the elastic behavior of a metal better than presently available lattice models. In particular, it allows for the effect of shearing stresses and volume conservation under loading, which the present lattice theory does not.

(2) It implicitly allows for nonuniform surface conditions characteristic of the polycrystalline metal surfaces on which many of the accommodation coefficient measurements have been carried out.

(3) Because the analysis directly involves the solution of partial differential equations, it tends to be simpler than the corresponding analysis of a discrete lattice model, and hence allows examination of more detailed effects, including surface temperature.

After these introductory remarks we now consider briefly the actual methods of analysis which have been used, and the results obtained.

The original continuum theory due to Landau [82] assumed a repulsive potential $V = \exp(-z/b)$, where z is the distance of the gas atom from the surface of the solid. The analysis used a perturbation procedure of the form outlined in Section 2. Thus the equation of motion of the gas atom is solved with the solid kept rigid; this solution is converted into a time-dependent force exerted by the gas atom at a point on the surface. From this force pulse the energy transferred to each mode of the solid is found as follows:

$$\Delta E(\omega) = \frac{4\pi^2 \omega^2 b^2 m_g^2}{N m_s} \; \mathrm{csch}^2 \frac{\pi \omega b}{u_{ni}} \tag{3.64}$$

where ω is the frequency of the mode, N is the number of atoms in the solid, u_{ni} is the incident velocity of the gas atom and m_g and m_s are the masses of the gas and lattice atoms respectively. The generalization of Eq. (3.64) for an initially moving oscillator is given in Eq. (3.86). An expression for the energy accommodation coefficient is then obtained [82] by integrating

Eq. (3.64) over the normal mode distribution of the elastic continuum. Gilbey [44] discusses the Landau theory, and points out that the Landau expression for the accommodation coefficient yields the high-temperature limit $\alpha \simeq 8\mu$, which is in error by a factor of two compared with the expected high-temperature limit, $\alpha \simeq 4\mu$. Gilbey shows that this error arises from the inability of the continuum to represent the higher-frequency motions of a lattice; in any case we would not expect the continuum model to be valid in this high-temperature limit. An approximate estimate of the energy accommodation coefficient based on a continuum model has also been given by Frenkel [83].

A more recent study using the continuum model is that of Karamcheti and Scott [84]. These authors give a direct comparison between the continuum and discrete lattice cases for corresponding conditions. They consider two different gas-surface interaction potentials, a truncated harmonic potential and a Morse potential, for which the corresponding lattice theories had already been developed. For the truncated harmonic potential they obtain an exact expression for the relative gas-atom-surface displacement as a function of time, and compare this with the corresponding quantity calculated by McCarroll and Ehrlich [63]. They find, as one would expect, that for a soft gas-surface interaction potential (i.e., long collision times) the results from the two models tend to agree. For the Morse potential Karamcheti and Scott use a procedure used by Landau and compare with the corresponding results of Trilling [73]. The calculations are made specifically for the system Ne/W at temperatures below room temperature, in order to compare with experimental data. For these conditions the accommodation coefficient calculated from the continuum model is larger than Trilling's results for the discrete lattice by a factor of 4 or 5; the potential parameters in this case have been adjusted so that the lattice model matches experiment. By altering the energy parameter D of the Morse potential reasonable agreement between the continuum theory and experiment can be obtained. We note that for both the discrete lattice and continuum theories the model is one-dimensional and the solid is initially at rest ($0°K$), so that the criticisms made in Section 3.2.2 apply here.

The most recent and probably the most comprehensive consideration of the continuum model is that of Trilling [81]. In this work the solid is a three-dimensional classical elastic solid defined by a Young's modulus and a Poisson ratio, or equivalently by two wave propagation velocities c_1 and c_2. A Morse-type potential is used, having a slightly different form from that given in Eq. (3.1):

$$V = 4D[\exp(-2r/b') - \exp(-r/b')]$$ (3.65)

where r is the gas-surface separation, b' is the range of the potential, and D is the well depth which now occurs at the separation given by $\exp(-r/b') = \frac{1}{2}$. The analysis uses the perturbation procedure discussed above, that is to say that the solid is assumed rigid for the calculation of the gas atom trajectory and the corresponding force pulse acting on the surface. The force pulse for the potential given by Eq. (3.65) is

$$F(t) = \frac{m_g u_{ni}^2}{b'} \frac{1 - \delta \cosh(u_{ni}t/b')}{\left[\cosh(u_{ni}t/b') - \delta\right]^2}$$ (3.66)

where

$$\delta \equiv \left[2D/\left(2D + m_g u_{ni}^2\right)\right]^{1/2}$$

The force pulse is then converted into a stress pulse by dividing the area by $\pi \ell_0^2$, where ℓ_0 is a characteristic length of the order of a lattice spacing. At this stage, it is assumed that the surface is effectively flat because only the normal component of stress is considered. Working from the solutions of the classical wave equation for the solid, Trilling then calculates the work done by the force $F(t)$ on the solid. This work done consists of two parts, the work on the solid due to the thermal motion of the surface, and the work done on the solid as a result of displacement induced by the inter-action.

Although the surface temperature is included at this stage of the analysis, most of the subsequent calculation of the accommodation coefficient assumes that the solid is initially at rest, i.e., $T_s = 0$. For this case Trilling gives a general integral expression for α, but considers limiting cases of this expression for limiting values of the parameters $\Lambda = (\ell_0/b'c_1)(2kT_g/m_g)^{1/2}$.

The parameter Λ is approximately the ratio of the Debye period in the solid to the collision time. As discussed at the beginning of this section the continuum approximation is most likely to be valid for $\Lambda \ll 1$, and the expression for α in this limit is

$$
\alpha = \left[1 - \exp\left(-u_{cr}^{\,2} \right) \right] + 24\,\mu \Lambda^3 \Psi(C)\, \mathcal{D}^{5/2} \Bigg\{ \left[\frac{5}{3\mathcal{D}} + \frac{4}{3\mathcal{D}} \right.
$$

$$
\left. + \left(1 + \frac{2}{3\mathcal{D}} \right) \frac{2\,u_{cr}}{\pi\,\mathcal{D}^{3/2}} + \frac{4\,u_{cr}^{\,2}}{3\,\mathcal{D}^2} + \frac{9\,u_{cr}^{\,3}}{8\,\pi\mathcal{D}^{5/2}} \right] \exp\left(-u_{cr}^{\,2} \right)
$$

$$
+ \frac{1}{2(\pi\mathcal{D})^{1/2}} \, \mathrm{erfc}\; u_{cr} \left[1 + \frac{2}{\mathcal{D}} + \frac{4}{3\,\mathcal{D}^2} \right] \Bigg\} \qquad (3.67)
$$

where $\mu = m_g/\rho \ell_0^{\,3}$ (ρ = density of solid), $\Psi(C)$ is a factor depending on the Poisson ratio, $\mathcal{D}$ is a dimensionless well depth D/kT_g, and u_{cr} is the critical gas velocity below which the gas atom is trapped on the surface. The velocity u_{cr} is given by the condition that the gas atom transfers its incident kinetic energy to the surface, and an expression for u_{cr} is given in Ref. [81].

Trilling compares Eq. (3.67) with a wide range of experimental data for the energy accommodation coefficient. The computation requires the selection of values for several physical constants. Some of these, such as the Poisson ratio and the Debye temperature are fairly accurately known. The model also includes a steric factor and the reference length ℓ_0. The most important selection process involves the range and well depth of the interaction potential. Good agreement with experiment is obtained using values for these two parameters which are similar to values used in connection with other models (see Sections 3.2 and 3.4). Some of the values used and the comparison between theory and experiment are reproduced in Table 3.2. The values marked with an asterisk in this table were calculated from Eq. (4.10) of Ref. [81] rather than from Eq. (3.67) above.

The continuum model is therefore capable of giving good agreement with the experimental data for α and clearly warrants further consideration. Two reservations must be made at this stage, however: (1) Using an expression which assumes $T_S = 0$ to compare with the experimental results is not a

TABLE 3.2

Comparison of Measured and Calculated Values of the Energy Accommodation
Coefficient of Inert Gases on Nonalkaline Metals (Trilling [81])

System	b', nm	D/k, °K	Λ	α (meas)	α (calc)
He/W	6.0	55	0.385	0.0164	0.0165*
Ne/W	6.0	200	0.20	0.043	0.0431
A/W	6.0	1000	0.14	0.26	0.25
Kr/W	6.9	1600	0.082	0.42	0.40
Xe/W	6.9	2560	0.070	0.66	0.65
He/Mo	6.0	50	0.36	0.022	0.023
Ne/Mo	6.0	200	0.16	0.045	0.045
A/Mo	6.0	1000	0.11	0.27	0.27
Kr/Mo	6.9	1600	0.066	0.43	0.43
Xe/Mo	6.9	2560	0.056	0.69	0.69
He/Al	6.0	50	0.35	0.073	0.066*
He/Ni	6.0	50	0.45	0.073	0.068
He/Pt	6.0	50	0.65	0.038	0.040
Ne/Fe	6.0	200	0.15	0.053	0.056
He/Be	6.0	50	0.127	0.145	0.141*
Ne/Be	6.0	200	0.065	0.315	0.316*

*See text, page 40.

satisfactory procedure, as has been discussed in Section 3.2.2. There is
some discussion of this point in Ref. [81]. (2) The low-temperature range
in which the continuum model is most likely to be valid is also the range
in which quantum effects are likely to be important.

3.4. SIMPLE CLASSICAL (CUBE) MODELS

During the first part of the 1960's a considerable amount of experimental data on the scattering of atomic beams from surfaces became available. The characteristic features of these scattering patterns, which we have already discussed briefly in Section 2, could not be explained in terms of the quantum mechanical and lattice theories available at that time. With a view to explaining the qualitative features of this data the so-called hard-cube model [85-86] was introduced. This model, although very simple, provided remarkably good agreement with the experimental scattering patterns. The hard-cube model has been superceded by the soft-cube model [87], in which some of the more extreme assumptions in the hard-cube model are eliminated, but with the penalty of increased complexity in the analysis. The analysis, as described in Section 3.4.2, is still relatively simple compared with, for example, the classical lattice theories.

3.4.1. The Hard-Cube Model

The hard-cube model is illustrated in Fig. 3.6, and the basic assumptions of the model are summarized as follows:

(1) Both the gas atoms and surface atoms are represented by rigid elastic particles (i.e., the intermolecular potential is impulsive).

(2) The gas-solid interaction potential is uniform in the plane of the surface, i.e., the surface is perfectly smooth and hence collisions with the surface do not change the tangential velocity component of the gas atom.

[Assumptions (1) and (2) are equivalent to assuming that the gas atoms behave as hard spheres, and the surface atoms as hard cubes oriented with one face parallel to the surface. Hence the name, hard-cube model.]

(3) The surface is represented by an ensemble of independent hard cubes confined by a square-well potential. This well serves to confine the atoms to the surface, but does not affect the motion of the gas atom.

(4) The surface atoms (i.e., hard cubes) have a Maxwellian velocity distribution. Only the velocity component normal to the surface is of interest here.

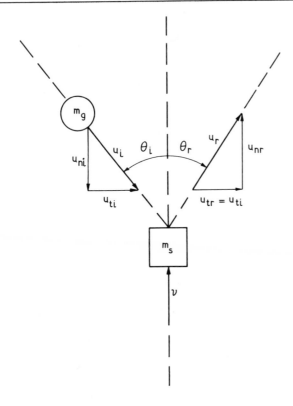

FIG. 3.6.   The hard–cube model.

With the exception of (2) these assumptions are the same as those used by Goodman [88] in calculations of the energy accommodation coefficient. An examination of the validity of these assumptions, together with those involved in the soft-cube model, is postponed until Section 3.4.4.

In Ref. [85] two methods of analysis of the model were described. The first was an approximate method based on mean speeds; this gave a closed-form result, but only for the position of the maximum of the scattering pattern. The second method yielded the full scattering pattern, but numerical integration was necessary. A more satisfactory analysis is that given in Ref. [86], which we outline here.

Referring to Fig. 3.6, the gas atom of mass m_g, velocity u_i, and angle
of incidence θ_i collides with the surface atom of mass m_s and velocity v_i,
and it is scattered with velocity u_r at angle θ_r. Since the tangential component u_{ti} is unchanged according to assumption (2), only the change in the normal component need be considered. It follows from assumption (1) than an
expression for u_{nr}, the normal velocity component of the scattered gas atom,
may be derived simply from the classical mechanics of rigid, elastic bodies:

$$u_{nr} = \frac{1 - \mu}{1 + \mu} u_{ni} + \frac{2}{1 + \mu} v_i \qquad (3.68)$$

where μ is the mass ratio m_g/m_s. With this knowledge of u_{nr} we may determine θ_r by using the following expression derived from the geometry
shown in Fig. 3.6:

$$\cot \theta_r = \frac{u_{nr}}{u_{ni}} \cot \theta_i \qquad (3.69)$$

It is clear at this stage that, according to the model, gas atoms which gain
energy in the collision are scattered at angles above the specular, whereas
those which lose energy are scattered at angles below the specular.

Since Eqs. (3.68) and (3.69) enable one to determine θ_r for any combination of u_{ni} and v_i, an expression for the scattering pattern (i.e., the
angular distribution) may be derived for any choice of distribution functions
for u_{ni} and v_i. We assume that the fraction of gas atoms having double (or
multiple) collisions with the surface is negligible; as discussed in Section
3.4.4 this restricts the validity of the analysis to small values of the mass
ratio μ. Under these conditions the differential rate at which collisions occur between incident gas atoms and surface atoms is given by

$$d^2R = V_R F(u_i)G(v_i)du_i dv_i \qquad (3.70)$$

where V_R is the relative velocity of the gas and surface atoms (along the surface normal) and F and G are the velocity distribution functions for gas atoms
and surface atoms, respectively. The probability distribution for scattering
is obtained by writing v_i as a function of u_i and θ_r [using Eqs. (3.68) and
(3.69)], integrating over u_i and dividing by the number of atoms striking the
surface per unit time $\overline{u_{ni}}$:

$$\frac{1}{\overline{u}_{ni}} \frac{dR}{d\theta_r} = \frac{1}{\overline{u}_{ni}} \int_{u_i = 0}^{\infty} V_R \, F(u_i) G(v_i) \left| \frac{dv_i}{d\theta_r} \right| du_i \tag{3.71}$$

From Eqs. (3.68) and (3.69)

$$v_i = \left(\frac{1 + \mu}{2} \sin \theta_i \cot \theta_r - \frac{1 - \mu}{2} \cos \theta_i \right) u_i \equiv B_1 u_i \tag{3.72}$$

Hence

$$\left| \frac{\partial v_i}{\partial \theta_r} \right| = \left(\frac{1 + \mu}{2} \sin \theta_i \csc^2 \theta_r \right) u_i \equiv B_2 u_i \tag{3.73}$$

The relative velocity is given by

$$V_R = u_i \cos \theta_i + v_i = (\cos \theta_i + B_1) u_i \tag{3.74}$$

The velocity distribution for the surface atoms in thermal equilibrium at temperature T_S is

$$G(v_i) dv_i = \left(\frac{m_s}{2 \pi k T_s} \right)^{1/2} \exp \left(-\frac{m_s v_i^2}{2 k T_s} \right) dv_i \tag{3.75}$$

The most important velocity distributions for the gas atoms are the Maxwellian and the monoenergetic.

First we consider a Maxwellian velocity distribution corresponding to an atomic beam from a source at temperature T_g:

$$F(u_i) du_i = \frac{4}{\sqrt{\pi}} u_i^2 \left(\frac{m_g}{2 k T_g} \right)^{3/2} \exp \left(-\frac{m_g u_i^2}{2 k T_g} \right) du_i \tag{3.76}$$

By substituting Eqs. (3.73)-(3.76) into Eq. (3.71) and integrating we obtain the scattering distribution

$$\frac{1}{\overline{u}_{ni}} \frac{dR}{d\theta_r} = \frac{3}{4} \left(\frac{m_s T_g}{m_g T_s} \right)^{1/2} \left(1 + \frac{m_s T_g}{m_g T_s} B_1^2 \right)^{-5/2} B_2 (1 + B_1 \sec \theta_i) \tag{3.77}$$

This equation gives the probability for scattering into unit angular range at angle θ_r. The scattering patterns given by this equation agree closely with

those given by the full numerical analysis in Ref. [85]. It should be noted
that Eq. (3.77) gives the angular distribution of the <u>flux</u> of gas atoms leaving
the surface; in cases where the experimental detector measures the density of
scattered atoms, a slightly modified expression should be used, as discussed
in Ref. [86]. According to Eq. (3.77) the scattering behavior depends only
on the parameters θ_i, μ, and m_gT_S/m_ST_g. We note that the quantity $(m_gT_S/m_ST_g)^{1/2}$ corresponds to the ratio of the mean speed of the solid atom to
that of the gas atom.

The second case to consider is that of a monoenergetic incident atomic
beam [89], i.e., the gas atoms all have incident velocity u_i. In this case
Eq. (3.71) becomes

$$\frac{1}{u_{ni}} \frac{dR}{d\theta_r} = \frac{1}{\sqrt{\pi}} \frac{u_i}{v_s} B_2 (1 + B_1 \sec \theta_i) \exp\left(-\frac{B_1^2 u_i^2}{v_s^2}\right) \tag{3.78}$$

where $v_s = (2kT_S/m_S)^{1/2}$ is the mean thermal speed of the surface atom.

Two examples of comparison between the hard-cube theory and experiment are shown in Figs. 3.7 and 3.8. The results in Fig. 3.7 show that the
hard-cube theory produces scattering patterns similar to those found experimentally, and gives the correct behavior for varying surface temperature at
fixed incident angle and gas temperature. As shown in Ref. [85] similar
agreement is obtained for the effects of varying incident angle θ_i and mass
ratio μ. For simple lobular scattering patterns, such as we are concerned
with here, the most important single parameter for use in comparing theory
and experiment is the position of the maximum of the lobe $\theta_{r\ max}$.

Figure 3.8 shows a plot of $\theta_{r\ max}$ against the combined parameter m_gT_S/m_ST_g for the experimental data of Saltsburg and Smith [90] for Ar/Ag and
the corresponding hard-cube theory. The agreement between theory and ex-
periment is again quite satisfactory. The theoretical results in Figure 3.8
were obtained from Eq. (3.77), and those in Fig. 3.7 from the full numerical integration described in Ref. [85]. In both these instances the experimental conditions correspond to a Maxwellian velocity distribution in the incident beam.

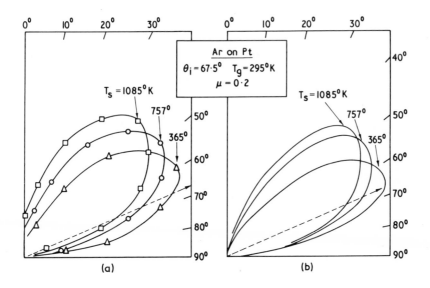

FIG. 3.7. Comparison between results from the hard-cube model and experimental results for the scattering of Ar/Pt. $\theta_i = 67.5°$, $T_g = 295°$ K (Logan and Stickney [85]).

In Refs. [85, 86] and [89] several other aspects of the hard-cube theory are discussed. These include the calculation of the velocity distribution $f_\theta(v)$ of the atoms scattered at angle θ_r, the effect of including an attractive potential in the interaction potential and the effect of surface roughness in the interaction potential. Some of these topics are covered by studies on the soft-cube model, described in following sections.

3.4.2.   The Soft-Cube Model

The soft-cube model is a logical successor to the hard-cube model and is still very simple, as illustrated in Fig. 3.9. The basic assumptions involved are:

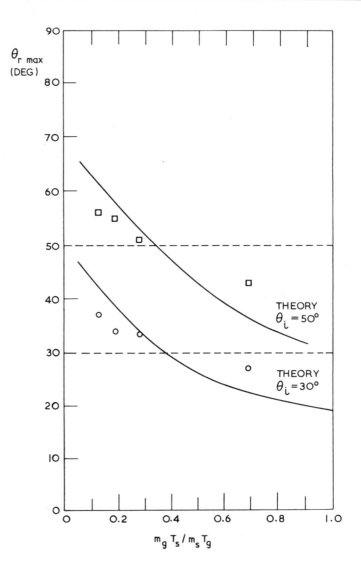

FIG. 3.8. Comparison between results from the hard-cube model and experimental results for the scattering of Ar/Ag. $T_S = 560°K$, $T_g = 300-1500°K$ (Logan et al. [86]).

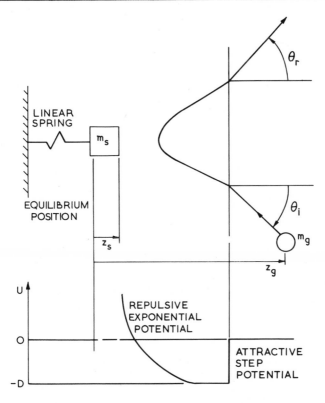

FIG. 3.9.   The soft–cube model.

(1)   The gas atom interacts with the surface through a potential which has two parts: (a) a stationary attractive part which increases the normal component of velocity of the gas atom before the repulsive collision and decreases it again afterwards; (b) an exponential repulsive part which acts between the gas atom and the single surface atom involved in the collision.

(2)   The gas–solid interaction potential is uniform in the plane of the surface, i.e., the surface is perfectly smooth, and hence collisions with the surface do not change the tangential velocity component of the gas atom.

[Assumptions (1) and (2) are to be compared with the corresponding assumptions made for the hard-cube model. It is these two assumptions which lead to the name "soft-cube" model in the present case.]

(3)   The surface atom involved in the collision is connected by a single spring to the remainder of the lattice, the latter being fixed.

(4)   The ensemble of oscillators which comprises the surface has an equilibrium distribution of energies at the temperature of the solid.

The question of the validity of these assumptions, together with those of the hard-cube model, is postponed until Section 3.4.4. We proceed now with the mathematical analysis of the model.

The method of analysis presented here follows that given in Ref. [87]. One slight change is that we do not include here the correction term in the force pulse which allows for the recoil of the surface atom. This term was included in the original analysis in an attempt to extend the range of validity to cases of high mass ratio ($\mu \simeq 1$), but due to the uncertainties involved in this term and the mathematical complication caused by it we choose here to leave it out and restrict the soft-cube model accordingly to $\mu \gtrsim 0.5$. (Details of the analysis of a modified soft-cube model with a truncated harmonic interaction potential have recently been published [91].)

The equations of motion of the gas and surface atoms are

$$\ddot{z}_g = \frac{F}{m_g} \tag{3.79}$$

$$\ddot{z}_s = -\frac{F}{m_s} + \omega^2 z_s \tag{3.80}$$

where z_g and z_s are the displacements shown in Fig. 3.9, F is the force, and ω is the natural frequency of the surface oscillator. For the repulsive part of the interaction we assume the form

$$U = U_0 \exp\left[-(z_g - z_s - z_{g0} - z_{s0})/b\right] = B \exp(-r/b) \tag{3.81}$$

where r is the relative coordinate $z_g - z_s$, B and b are constants of the potential, and the subscript 0 denotes quantities at the turning point (i.e., at minimum r). The analysis of the soft-cube model is basically a perturbation procedure. There are two obvious zero-order situations we can consider in connection with the coupled equations of motion (3.79) and (3.80). (a) We may fix the surface atom rigidly in position, and consider the collision of the gas atom with this fixed body. This corresponds to the perturbation procedure as already discussed in Section 2. (b) We may remove the spring, i.e., set $\omega = 0$, and consider the collision between two free particles m_g and m_s. We discuss the relative merits of these procedures in Section 3.4.4, but for the present analysis we use procedure (b).

The zero-order force pulse in this case has the following form

$$F = (\pi^2 m_r b\omega^2 / 2\gamma^2) \, \text{sech}^2 \, \tau^* \tag{3.82}$$

where

$$m_r \equiv \frac{m_g m_s}{m_g + m_s}$$

$$\tau^* \equiv \frac{t - t_0}{t_c}$$

$$t_c \equiv \left(\frac{2 \, m_r b^2}{U_0}\right)^{1/2}$$

and

$$\gamma = \frac{1}{2} \, \pi\omega t_c$$

For the perturbation procedure (b) the potential energy at the turning point U_0 is fully determined, and so also are the collision time t_c and γ. In this analysis of the soft-cube model, however, we choose to retain γ as a free parameter, the value of which is chosen by requiring that the impulse applied to the gas atom leads to the correct outgoing angle, as follows. We define a dimensionless impulse:

$$I = \frac{1}{m_g u_{ni}'} \int_{-\infty}^{\infty} F \, dt \tag{3.83}$$

where u'_{ni} is the incident velocity of the gas atom (normal component) within the potential well. Substituting for F from Eq. (3.82) in (3.83) and integrating we obtain

$$I = \frac{2\pi}{1+\mu} \frac{\sigma}{\gamma} \tag{3.84}$$

where $\sigma \equiv \omega b / u_{ni}'$.

By using the condition that the tangential component of velocity of the gas atom is unchanged, the impulse I can also be related to the outgoing angle θ_r of the gas atom:

$$I = 1 + \frac{u_{nr}'}{u_{ni}'}$$

$$= 1 + \left[\frac{\left(\cot^2 \theta_r / \cot^2 \theta_i \right) + D^*}{1 + D^*} \right]^{1/2} \tag{3.85}$$

where $D^* \equiv 2 D / m_g u_{ni}^2$.

From Eqs. (3.84) and (3.85) we can solve for γ as a function of the outgoing angle θ_r. In applying this condition we are effectively saying that our force pulse is a function of the momentum transfer (impulse) of the collision. To this extent the treatment goes beyond the usual straightforward perturbation procedure.

We now consider the determination of θ_r for given initial conditions. This may be done by applying the condition of energy conservation; i.e., the net energy lost by the gas atom in the collision must equal the net energy gained by the surface oscillator. The energy transferred to the oscillator in the collision is given by [87]

$$\Delta E_s = \mu \frac{m_g u_{ni}'^2}{2} \left(J^2 + \frac{2 J}{\mu} \frac{v \cos \omega t_0}{u_{ni}'} \right) \tag{3.86}$$

where

$$J = \frac{1}{m_g u_{ni}'} \int_{-\infty}^{\infty} \exp \left[i\omega(t - t_0) \right] F(t) \, dt \tag{3.87}$$

Using Eq. (3.82) in Eq. (3.87) we obtain

$$J = (\gamma \operatorname{csch} \gamma) I \qquad (3.88)$$

where I is given by Eq. (3.84). Once again using the assumption that the tangential component of velocity of the gas atom is unchanged, the energy loss of the gas atom may be written

$$\Delta E_g = \frac{m_g u_{ni}^2}{2} \left(1 - \frac{\cot^2 \theta_r}{\cot^2 \theta_i} \right) \qquad (3.89)$$

Equating the right-hand sides of Eqs. (3.86) and (3.89) and solving for
$(v \cos \omega t_0)/u_{ni}'$:

$$v_c \equiv \frac{v \cos \omega t_0}{u_{ni}'}$$

$$= \frac{1}{2 J} \left[\left(\frac{u_{ni}}{u_{ni}'} \right)^2 \left(1 - \frac{\cot^2 \theta_r}{\cot^2 \theta_i} \right) - \mu J^2 \right] \qquad (3.90)$$

Since J is a function of γ, and γ is a function of θ_r, v_c is a function of θ_r. We note that in the limit $\gamma = 0$, $\gamma \operatorname{csch} \gamma = 1$ and the result corresponds to the impulsive collision between two free particles, as in the hard-cube model. In the limit $\gamma \to \infty$, $\gamma \operatorname{csch} \gamma \to 0$ and the collision is elastic, with $\theta_r = \theta_i$.

We have shown above that there is a one-to-one relation between the undisturbed velocity of the oscillator at the turning point $v \cos \omega t_0$ and θ_r. To find the angular distribution of the outgoing gas atoms, we therefore need to know the probability distribution for $v \cos \omega t_0$. The velocity amplitude v is simply related to the energy of the oscillator, and the distribution of energies is known as a function of the equilibrium temperature of the solid.

To find the distribution of phase angles ωt_0 at the turning point, or phase probability, we proceed as follows. Integrating the equation of motion of the gas atom [Eq. (3.79)] twice gives

$$z_g - z_{g0} = -u_{ni}'(t - t_0) - \frac{1}{m_g} \int_t^{t_0} \int_{-\infty}^{t_1} F(t_2) \, dt_2 dt_1 \qquad (3.91)$$

where t_1 and t_2 are dummy variables of integration. Since F has a $sech^2$ form and cuts off rapidly for $|t_0 - t| > t_c$, we may extend the lower limit of the t_1-integral to $-\infty$ for cases where $t \ll t_0 - t_c$. Then Eq. (3.91) becomes

$$z_g = z_{g0} - z_{gf} - u_{ni}'(t - t_0); \quad t \ll t_0 - t_c \tag{3.92}$$

where

$$z_{gf} \equiv \frac{1}{m_g} \int_{-\infty}^{t_0} \int_{-\infty}^{t_1} F(t_2) \, dt_2 dt_1 \tag{3.93}$$

The physical interpretation of Eq. (3.92) is shown in Fig. 3.10; this shows that if the gas atom had proceeded on its undisturbed trajectory, it would have been at a position $z_{g0} - z_{gf}$ at time t_0. It is clearly a reasonable assumption, however, that there is a uniform distribution of undisturbed trajectories; i.e., the time instants t_i at which the undisturbed trajectories would cross the line $z_g = 0$ are uniformly distributed. From Fig. 3.10 it is seen that t_i is given by

$$t_i = t_0 + \frac{z_{g0} - z_{gf}}{u_{ni}'}$$

Hence the probability that in a collision the phase angle will be in the range ωt_0 to $\omega t_0 + d(\omega t_0)$ is

$$P(\omega t_0) d\omega t_0 = \frac{1}{2\pi} \frac{d\omega t_i}{d\omega t_0} d\omega t_0$$

$$= \frac{1}{2\pi} \left[1 + \sigma \frac{d}{d\omega t_0} \left(\frac{z_{g0} - z_{gf}}{b} \right) \right] d\omega t_0 \tag{3.94}$$

Carrying out the differentiation as described in Ref. [87] we obtain

$$P(\omega t_0) = \frac{1}{2\pi} \left(1 - v_c + \sigma v_s \frac{d\theta_r}{dv_c} \frac{d\Gamma}{d\theta_r} \right) \tag{3.95}$$

where $v_c \equiv (v \cos \omega t_0)/u_{ni}'$, $v_s \equiv (v \sin \omega t_0)/u_{ni}'$ and Γ is some function of θ_r. By using Eq. (3.90) we can in principle write $d\theta_r/dv_c$ as a function of θ_r.

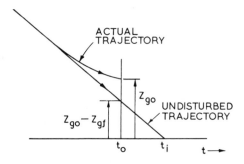

FIG. 3.10. Trajectory of incident gas atom.

We now consider the scattering of a monoenergetic beam from the surface. The use of a monoenergetic beam allows a closed-form expression
for the scattering distribution to be obtained, and many of the modern experiments actually use approximately monoenergetic beams. Furthermore
we expect the scattering patterns for Maxwellian and monoenergetic beams
to be closely similar when the well depth D/k is large compared to the temperature of the incident beam. For unit beam density the differential rate
at which collisions occur between incident gas atoms and surface atoms is
given by

$$d^2R = u_{ni}{}'P_1(\omega t_0)P_2(v)d\omega t_0\,dv \tag{3.96}$$

where $P_1(\omega t_0)$ is the phase probability given by Eq. (3.95) and $P_2(v)$ is the
probability distribution for the velocity amplitude of the oscillators given by

$$P_2(v)dv = \frac{m_s v}{kT_s}\exp\left(-\frac{m_s v^2}{2\,kT_s}\right)dv \tag{3.97}$$

By substituting Eq. (3.97) into (3.96) and transforming to the variables v_c
and v_s we obtain

$$d^2R = u_{ni}{}'\,\frac{m_s u_{ni}'^2}{kT_s}\,P_1(v_c,v_s)\exp\left[-\frac{m_s u_{ni}'^2}{2kT_s}(v_c{}^2 + v_s{}^2)\right]dv_c dv_s \tag{3.98}$$

The probability for scattering into the angular range θ_r to $\theta_r + d\theta_r$ is then given by

$$\frac{1}{u_{ni}} \cdot \frac{dR}{d\theta_r} = \frac{m_s u_{ni}'^2}{kT_s} \int_{-\infty}^{\infty} P_1(v_c, v_s) \exp\left[-\frac{m_s u_{ni}'^2}{2kT_s} (v_c^2 + v_s^2) \right] \left| \frac{\partial v_c}{\partial \theta_r} \right|_{v_s} dv_s \quad (3.99)$$

In Eq. (3.90) we have shown that v_c is a function of θ_r and not v_s. Hence terms involving v_c and $\left| \partial v_c / \partial \theta_r \right|_{v_s}$ may be brought outside the integral in Eq. (3.99). We also note that the antisymmetric v_s term in the phase probability Eq. (3.95) will give zero contribution when integrated over the range $-\infty \to \infty$. The integral over v_s then becomes a standard definite integral, and the scattering distribution is

$$\frac{1}{u_{ni}} \cdot \frac{dR}{d\theta_r} = \left(\frac{m_s u_{ni}'^2}{2\pi kT_s} \right)^{1/2} (1 - v_c) \left| \frac{dv_c}{d\theta_r} \right| \exp\left(\frac{-m_s u_{ni}'^2 v_c^2}{2kT_s} \right) \quad (3.100)$$

The above scattering distribution is expressed in terms of the variable v_c; but we know v_c as a function of γ [Eq. (3.90)], and we know θ_r as a function of γ [Eqs. (3.84) and (3.85)]. Hence, using γ as the working variable we may calculate the scattering distribution as a function of θ_r. The derivative $dv_c/d\theta_r$ is obtained from Eq. (3.90).

Equation (3.100) is the scattering pattern given by the basic two-dimensional soft-cube model. In Ref. [87] an additional factor was included in an attempt to allow for the three-dimensional nature of the detector in a typical gas-surface scattering experiment. Since this factor does not form a fundamental part of the model, and is in any case open to some uncertainty, we do not reproduce it here.

We now consider comparison between the results from the soft-cube model and experiment. For specified experimental conditions, the results from the soft-cube model are governed by the values of the quantities ω, b, and D. It may be seen from the above analysis [Eq. (3.84)] that the first two of these occur only in the combination $\sigma = \omega b / u$. It is therefore convenient to define a quantity n such that

$$n \equiv \frac{b}{a_0} \frac{\hbar\omega}{k\Theta_D} \quad (3.101)$$

where Θ_D is the Debye temperature of the solid and a_0 is the Bohr radius. If the correct frequency to use for the single oscillator in the soft-cube model is given approximately by the Debye frequency, then we expect that the quantity n has the value $n \simeq b/a_0$; by comparison with values for the range of the potential found in gas-gas collisions, we expect this quantity to have a value of about 0.5. There is evidence [92, 93] that the surface layer of a solid has a somewhat lower Debye temperature (i.e., a lower maximum frequency) than the bulk of the solid, and thus values of n less than 0.5 might be expected.

In Ref. [87] a procedure was described for obtaining a best fit between theory and experiment for a series of results on a given gas/surface system. The comparison procedure was based on the angular position of the maximum of the distribution. Some values of n and D obtained from this procedure are shown in Table 3.3. In this table values are given for b obtained from the values of n by using the assumption that $\omega = k\Theta_D/\hbar$. It is seen that these values for b are of the same order as the literature values for the range of the repulsive exponential potential in the corresponding gas-gas collision. Similarly the values of D are comparable with literature values of the heat of adsorption, where these are available. Thus for reasonable values of the parameters n and D the soft-cube model gives good agreement with experimental data for the position of the maximum of the scattering pattern in several gas/surface systems.

A specific example of comparison between the soft-cube model and experimental data from Ref. [90] is shown in Fig. 3.11 for the system Ar/Ag, where the n and D values from Table 3.3 have been used to calculate the full in-plane distribution. It is seen that the distribution from the soft-cube model is narrower than the experimental distribution. There are several reasons why a discrepancy of this type might be expected, as follows:

(1) The experimental measurements include gas atoms which are adsorbed on the surface and then reemitted. In Fig. 3.11 we have corrected the experimental results for the presence of these reemitted atoms by assuming that the flux along the normal is due entirely to reemitted atoms, and that these atoms have a diffuse distribution such that the reemitted flux at angle θ_r is proportional to $\cos \theta_r$. Subtracting

TABLE 3.3

Comparison of Values Used in the Soft-Cube Model with
Literature Values (Logan and Keck [87])

Gas	Surface	n	b,[b] nm	D/k, °K	b,[d] nm	ΔH/k,[g] °K
			Soft-cube model[a]		Literature values	
He	Au	0.50[c]	2.6	50[c]	2.6[e]	50[h]
Ne	Au	0.46	2.4	300	2.7[f]	200[i]
Ar	Au	0.65	3.4	1200	3.4[f]	950[i]
	Ag	0.35	1.8	600		
	Pt	0.30	1.6	1600		
Xe	Au	0.76	4.0	6000	4.0[f]	4500[i]
	Ag	0.39	2.1	2000		

[a]Values for gas-surface interactions obtained by comparison of experiment with the soft-cube model.

[b]Values for the interaction range obtained from the values of n by assuming $\omega = k\Theta_D/\hbar$ in Eq. (3.101).

[c]Obtained by comparison with literature values.

[d]Gas-gas interaction.

[e]Derived from value estimated by Goodman [67] for the range of a Morse potential in the gas-gas collision.

[f]Derived from values for the range of a Morse potential for the gas-gas collision, determined by Konowalow and Hirschfelder [94] by comparison with experiments.

[g]Gas-surface interaction with W.

[h]Values estimated by Trilling [73].

[i]Experimental values on tungsten given by Ehrlich [95].

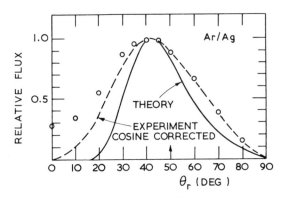

FIG. 3.11. In-plane scattering distributions for Ar/Ag; comparison between results from soft-cube model, using values for n and D given in Table 3.3, experimental data (o) and same experimental data after correction for an assumed cosine component (Logan and Keck [87]).

this component from the experimental results and renormalizing leads to the distribution shown by a broken line in Fig. 3.11.

(2) The experimental results in Fig. 3.11 were obtained with an incident Maxwellian beam, whereas Eq. (3.100) applies for a monoenergetic beam. The use of a Maxwellian beam in the soft-cube model would give a somewhat broader distribution.

(3) In deriving Eq. (3.100) effects of surface structure, which would tend to broaden the distribution, have been neglected. We return to a discussion of the surface structure in Section 3.4.4.

3.4.3. The Soft-Cube Model and the Energy Accommodation Coefficient

In addition to the scattering distributions discussed in the previous section, the soft-cube model has also been used for the calculation of the energy accommodation coefficient [80]. It is notable in this respect because it is a model based on classical mechanics which includes surface temperature effects, in contrast to the classical lattice models as discussed in Section 3.2. In fact the soft-cube theory enables quite good agreement with the experimental data to be obtained, for physically reasonable values of the parameters b and D.

The basis of the analysis is the same as that given in Section 3.4.2. Eq. (3.90) relating the initial velocity of the oscillator to the outgoing state of the gas atom can be written.

$$v_c \equiv \frac{v \cos \omega t_0}{u_{ni}{}'} = \frac{1}{2J} \left(\frac{\mathscr{E}_i - \mathscr{E}_r}{\mathscr{E}_i + \mathscr{D}} - \mu J^2 \right) \tag{3.102}$$

where $\mathscr{E}_i \equiv m_g u_{ni}{}^2/2kT_g$ and $\mathscr{E}_r \equiv m_g u_{nr}{}^2/2kT_g$ are, respectively, the dimensionless energies of the gas atom associated with the incident and outgoing normal components of velocity (outside the potential well), v is the velocity amplitude of the oscillator, $\mathscr{D}$ is the dimensionless well depth D/kT_g, and J is given by Eq. (3.87). Introducing the probability distributions for the phase ωt_0 and the velocity v, the probability that a gas atom has energy $\mathscr{E}_r > \mathscr{E}^*$ after collision with the oscillator is given by [87]

$$W(\mathscr{E}^*, \mathscr{E}_i) = \frac{1}{2} \left\{ 1 + \mathrm{erf} \left[\left(\frac{m_s u_{ni}{}'^2}{2kT_s} \right)^{1/2} v_c{}^* \right] \right.$$

$$\left. + \frac{1}{\sqrt{\pi}} \left(\frac{2kT_s}{m_s u_{ni}{}'^2} \right)^{1/2} \exp \left(- \frac{m_s u_{ni}{}'^2 v_c{}^{*2}}{2kT_s} \right) \right\} \tag{3.103}$$

where $v_c{}^*$ is given by Eq. (3.102) with $\mathscr{E}_r = \mathscr{E}^*$. Assuming unit incident flux of gas atoms, $W(\mathscr{E}^*, \mathscr{E}_i)$ also represents the flux of gas atoms leaving with energy $\mathscr{E}_r > \mathscr{E}^*$.

In the analysis [87] leading to Eq. (3.103) the total scattered flux was normalized to unity for integration over the range $-\infty < v_c < \infty$; note from Eq. (3.103) that $W(\mathscr{E}^*, \mathscr{E}_i) \to 1$ as $v_c{}^* \to \infty$. Strictly, this analysis is only valid for $-\infty < v_c < 1$ because $v_c > 1$ implies that the surface atom is moving away from the gas atom at the turning point, which is not realistic; furthermore, the expression for the phase probability breaks down under these conditions. Extending the range of v_c to ∞ introduces negligible error in most cases because the scattering probability decreases exponentially with v_c^2, but the incorrect normalization can cause small errors in the value of $W(\mathscr{E}^*, \mathscr{E}_i)$ and hence quite large percentage errors in the sticking probability $1 - W(0, \mathscr{E}_i)$ for cases where $W(0, \mathscr{E}_i) \simeq 1$. The effect of these errors on the calculation of the accommodation coefficient seems to be negligible except for cases involving He (and to a lesser extent Ne) where values of $W(0, \mathscr{E}_i) \simeq 1$ (i.e., very low sticking probability) are likely to occur. For completeness, however, we replace Eq. (3.103) by an expression corresponding to normalization over the range $- \infty < v_c < 1$ as follows:

$$
W(\mathscr{E}^*, \mathscr{E}_i) = \frac{1}{2} \left\{ 1 + \mathrm{erf} \left[\left(\frac{m_s u_{ni}'^2}{2kT_s} \right)^{1/2} v_c^* \right] \right.
$$

$$
\left. + \frac{1}{\sqrt{\pi}} \left(\frac{2kT_s}{m_s u_{ni}'^2} \right)^{1/2} \exp \left(- \frac{m_s u_{ni}'^2 v_c^{*\,2}}{2kT_s} \right) \right\}
$$

$$
\times [W(1)]^{-1}, \qquad v_c^* \leq 1
$$

$$
= 1, \qquad v_c^* > 1 \qquad\qquad (3.104)
$$

where $W(1)$ is given by the right-hand side of Eq. (3.103) with $v_c^* = 1$. In this case we note that $W(\mathscr{E}^*, \mathscr{E}_i) = 1$ at $v_c^* = 1$.

In order to calculate the accommodation coefficient we need to calculate the energy flux leaving the surface. The energy flux may be considered as consisting of three parts. First, a fraction of the gas atoms may be temporarily trapped on the surface, i.e., $\mathscr{E}_r < 0$ after collision with the oscillator. We assume that these atoms come into equilibrium with the surface before eventually returning to the gas with, therefore, an average energy $2kT_s$ per atom. Second, there is an energy flux associated with the tangential component of velocity of those gas atoms which are not trapped on the surface. Since the tangential velocity is assumed to remain unchanged in the soft-cube model the average energy per atom associated with this tangential component is kT_g, the same as the incident value. Finally, there is the energy flux associated with the outgoing normal component of velocity of the gas atoms which are not trapped. This last energy flux depends on the details of the interaction and, as is shown below, can be expressed in terms of the quantity $W(\mathscr{E}_r, \mathscr{E}_i)$.

It is convenient to carry out separate calculations of the energy accommodation coefficient for the two limiting cases $|T_s - T_g| \gg T_g$ and $|T_s - T_g| \ll T_g$, as follows:

(a) $|T_s - T_g| \gg T_g$. For the incident Maxwellian gas the flux of atoms crossing a surface with energy (associated with the component of velocity normal to the surface) in the range $\mathscr{E}_i$ to $\mathscr{E}_i + d\mathscr{E}_i$ is given by

$$P(\mathscr{E}_i)d\mathscr{E}_i = e^{-\mathscr{E}_i}d\mathscr{E}_i \qquad (3.105)$$

Integrating over the energy of the incident gas atoms the energy flux leaving the surface is given by [80]

$$\mathscr{E}_c = \int_0^\infty \int_0^\infty W(\mathscr{E}_r, \mathscr{E}_i)d\mathscr{E}_r e^{-\mathscr{E}_i}d\mathscr{E}_i \qquad (3.106)$$

The fraction of gas atoms f_t trapped on the surface is given by the fraction which are "scattered" with $\mathscr{E}_r < 0$, i.e.,

$$f_t = \int_0^\infty [1 - W(0, \mathscr{E}_i)]e^{-\mathscr{E}_i}d\mathscr{E}_i \qquad (3.107)$$

Since these atoms are assumed to come into equilibrium with the surface the energy flux due to the trapped atoms is $2(T_s/T_g)f_t$ (in units of kT_g). It then simply follows [80] that the energy accommodation coefficient defined by Eq. (2.2) is given by

$$\alpha = \frac{2(T_s/T_g)f_t + \mathscr{E}_c - (1 + f_t)}{2[(T_s/T_g) - 1]} \qquad (3.108)$$

(b) $|T_s - T_g| \ll T_g$. As pointed out in Section 3.1.1. the definition of the accommodation coefficient in Eq. (2.2) is often taken in the limit $T_g \to T_s$, and many of the experiments [79] are conducted with $|T_s - T_g|$ small so as to approximate to this limit. Due to the mathematical approximations used in the analysis [87] of the soft-cube model, taking the limit $T_g \to T_s$ in Eq. (3.108) is not satisfactory because the numerator does not necessarily go exactly to zero at $T_s = T_g$. A suitable expression which is well behaved at $T_s = T_g$ can be obtained, however, by considering a different form of the definition of the accommodation coefficient.

Let $\mathscr{E}_N \equiv \mathscr{E}_r - \mathscr{E}_i$ denote the net energy transfer per atom from the surface to the gas. In general $\mathscr{E}_N$ can be expanded in a series as follows:

$$\mathscr{E}_N(T_s, T_g) = \mathscr{E}_N\bigg|_{T_s = T_g} + \frac{\partial \mathscr{E}_N}{\partial T_s}\bigg|_{T_s = T_g} (T_s - T_g)$$

$$+ \frac{1}{2!} \frac{\partial^2 \mathscr{E}_N}{\partial T_s^2}\bigg|_{T_s = T_g} (T_s - T_g)^2 + \cdots \tag{3.109}$$

The first term on the right-hand side must be zero for thermal equilibrium, and hence

$$\mathscr{E}_N \bigg|_{\lim T_g \to T_s} = \frac{\partial \mathscr{E}_N}{\partial T_s}\bigg|_{T_s = T_g} (T_s - T_g) \tag{3.110}$$

Noting that $\mathscr{E}_N$ is defined in units of kT_g we obtain from Eq. (3.110) and Eq. (2.2), in the limit $T_g \to T_s$

$$\alpha_{\lim} = \frac{T_g}{2} \frac{\partial \mathscr{E}_N}{\partial T_s}\bigg|_{T_s = T_g} \tag{3.111}$$

$\mathscr{E}_N$ is given by the numerator in Eq. (3.108); using this in Eq. (3.111) we obtain [80]

$$\alpha_{\lim} = \frac{T_g}{2}\left(\frac{2}{T_g} f_t + \frac{\partial \mathscr{E}_c}{\partial T_s} + \frac{\partial f_t}{\partial T_s}\right)\bigg|_{T_s = T_g}$$

$$= \frac{T_g}{2} \int_0^\infty \Bigg\{ \frac{2}{T_g} [1 - W(0, \mathscr{E}_i)]$$

$$+ \left[\int_0^\infty \frac{\partial W(\mathscr{E}_r, \mathscr{E}_i)}{\partial T_s} d\mathscr{E}_r\right] - \frac{\partial W(0, \mathscr{E}_i)}{\partial T_s} \Bigg\} e^{-\mathscr{E}_i} d\mathscr{E}_i \tag{3.112}$$

An explicit expression for the derivative $\partial W(\mathscr{E}_r, \mathscr{E}_i)/\partial T_s$ is given in Ref. [80].

We now compare the preceeding theoretical expressions [Eqs. (3.108) and (3.112)] with experiment. We choose the experimental results for the rare gases on tungsten, as represented by the results of Watt and Moreton [96] for the systems He/W and Ar/W, and of Thomas et al. [79] for He, Ne, Ar, Kr and Xe/W. The results in Ref. [79] have already been introduced in Section 3.2.2 for comparison with the classical lattice theory. The experimental conditions used by Watt and Moreton approximate the condition $|T_S - T_g| \gg T_g$, and therefore their results are compared with Eq. (3.108), while the conditions of Thomas et al. approximate to the condition $|T_S - T_g| \ll T_g$, requiring the use of Eq. (3.112).

Some examples of the comparison between theory and experiment are shown in Figs. 3.12 and 3.13, reproduced from Ref. [80]. Further cases are given in Ref. [80]. Values have been selected for the parameters n [Eq. (3.101)] and D to provide good agreement with experiment; the actual values are shown in the figure captions. In choosing a value of D the literature values of the relevant heats of adsorption on tungsten were used as a guide. We also note that the values of the range parameter b obtained from the value of n by assuming $\omega = k\Theta_D/\hbar$ in Eq. (3.101) are in reasonable agreement with the gas-gas values, as was the case in Table 3.3. A complete table similar to Table 3.3 is given in Ref. [80]; it is not reproduced here.

From Fig. 3.12 we see that the theory provides good agreement with the experimental results of Watt and Moreton [96] for the chosen values of n and D. The theory does predict that α is approximately independent of T_S under these conditions, a feature which may be generally true as already mentioned in Section 3.2.2. From Fig. 3.13 we see that the theory also provides good agreement with the experimental results of Thomas et al. [79] for Ne/W and He/W, using the chosen values of n and D. An important feature here is that the soft-cube theory does predict the experimentally observed minimum in the value of α. The classical lattice theories also predict this minimum but, as already discussed in Section 3.2, they do not include surface temperature effects.

In the course of calculating α from Eqs. (3.108) and (3.112) one calculates the quantity f_t, i.e., the fraction of the incident gas atoms which are

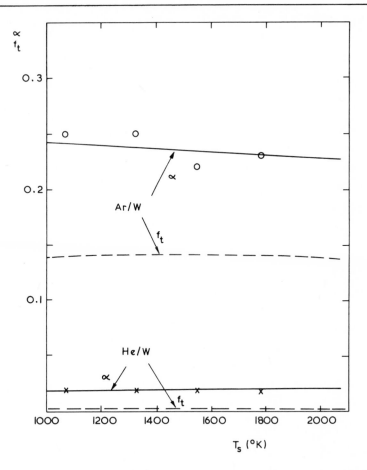

FIG. 3.12. Theoretical results from soft–cube model for α [Eq. (3.108)] and f_t [Eq. (3.107)]. Corresponding experimental points for α from Ref. [96]. $T_g = 300°$K. For He/W, $n = 0.33$, $D/k = 50°$K. For Ar/W, $n = 0.285$, $D/k = 950°$K (Logan [80]).

initially trapped on the surface. This quantity is plotted alongside α in Figs. 3.12 and 3.13. It is seen that there is, as to be expected, a strong correlation between f_t and α, particularly at lower temperatures. From Fig. 3.13 it seems likely that the increase in α at low temperatures is related to the increase in f_t. This is the reason why $\alpha \to 1$ as $T \to 0$. Even at higher

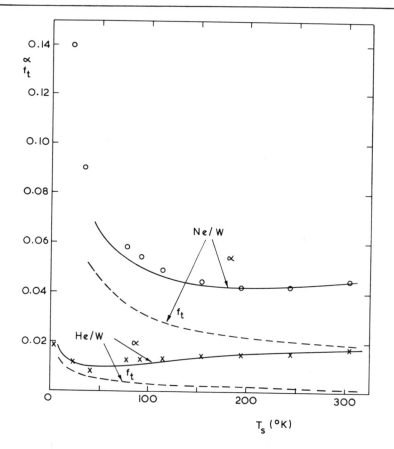

FIG. 3.13. Theoretical results from soft-cube model for α [Eq. (3.112)] and f_t [Eq. (3.107)]. Corresponding experimental points for α from Ref. [79]. For He/W, $n = 0.33$, $D/k = 50°$K. For Ne/W, $n = 0.335$, $D/k = 350°$K (Logan [80]).

temperatures, f_t plays an important role; for the case of Ar/W in Fig. 3.12 f_t is only about 0.15, but this small fraction of the atoms accounts for more than half the net energy exchange with the surface; this is because the surface is much hotter than the gas and we have assumed that atoms which are trapped come into equilibrium with the surface before leaving it.

3.4.4. Discussion of Assumptions in the Cube Models

The assumption which is common to both the hard- and soft-cube models is that the surface is effectively flat, i.e., the gas–solid interaction potential is uniform in the plane of the surface. We now consider in detail the likely validity of this important assumption. One obvious reason for making this assumption, although not a justification, is that it greatly simplifies the calculation of the scattering patterns. This is illustrated by Eqs. (3.68) and (3.69) in the analysis of the hard-cube model, where a simple expression is obtained relating the incident and outgoing angles of the gas atom.

Figure 3.14 shows a typical gas–surface interaction potential calculated [97] by summing the pairwise interactions between the gas atom and a square array of surface atoms. A Lennard–Jones 6-12 potential [Eq. (3.119)] was actually used for this calculation, but here we are only concerned with the qualitative form of the resulting potential contours, which would be the same for any reasonable pairwise interaction potential. The parameter β labelling the equipotential contours indicates the ratio of the total potential energy on that contour to the binary binding energy of the Lennard–Jones potential. A gas atom approaching the surface (normally) with high kinetic energy (at infinity) penetrates to contours with $\beta \gg 1$ where the surface appears to the gas atom as an array of individual approximately spherical atoms. In this case the scattering behavior is determined by the structure of the equipotential contours, and this is widely referred to as "structure scattering". Theories dealing specifically with structure scattering are discussed in Section 3.5. On the other hand a gas atom approaching the surface with low kinetic energy would penetrate only about as far as the contour $\beta = 0$. It is seen that over most of the surface this gas atom encounters contours which are approximately flat. If no other effects were involved the scattering of an atom with energy in this range (thermal energy) would give a scattering pattern corresponding approximately to that from an ideal flat surface. For gas atoms with this energy, however, inelastic collisions (creation or annihilation of phonons in the solid) have a significant effect on the energy, and hence trajectory, of the gas atoms. In particular, if the temperature of the solid is high relative to the effective temperature of the gas atoms, the scattering mechanism of prime importance is the thermal

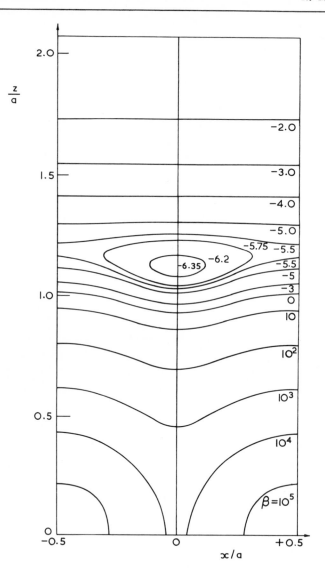

FIG. 3.14. Equipotential contours for typical gas–surface interaction
potential (McClure [97]).

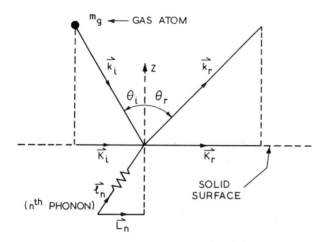

FIG. 3.15. An N-phonon inelastic scattering process about the diffraction peak denoted by $\underline{G}$. For simplicity in this figure all vectors are drawn in the plane containing $\underline{k}_i$ and the surface normal (Goodman [98]).

motion of the surface atoms, in the direction normal to the surface. Scattering in this regime is widely referred to as "thermal scattering," and it is in this regime that we might expect the flat-surface assumption in the cube models to be a valid approximation.

The considerations in the preceding paragraph were based on the classical mechanics of the interaction. A quantum mechanical justification for the flat-surface assumption in the cube models has been given by Goodman [98]. The situation is illustrated in Fig. 3.15. A gas atom with incident wave vector $\underline{k}_i$ interacts with the surface and is scattered with wave vector $\underline{k}_r$. For simplicity the vectors in Fig. 3.15 are drawn in the plane containing $\underline{k}_i$ and the surface normal. For a general N-phonon interaction the momentum and energy conservation conditions are [98]:

$$\underline{K}_r = \underline{K}_i + \underline{G} + \sum_{n=1}^{N} (\pm \underline{L}_n) \tag{3.113}$$

$$\frac{\hbar^2 k_r^2}{2\,m_g} = \frac{\hbar^2 k_i^2}{2\,m_g} - \sum_{n=1}^{N} (\pm\hbar\omega_n) \tag{3.114}$$

where $\underline{K}_i$ and $\underline{K}_r$ are the components in the surface plane of $\underline{k}_i$ and $\underline{k}_r$, $\underline{G}$ is a surface reciprocal lattice vector, and $\underline{L}_n$ is the tangential component of $\underline{\ell}_n$, the nth phonon of frequency ω_n. The significant result shown by Goodman is that, for the same energy, phonon momenta $\hbar\ell_n$ are in general considerably smaller than gas atom momenta $\hbar k$. The ratio of the momenta is given by

$$\ell_n/k = \frac{1}{2}\,u/c \tag{3.115}$$

where u is the gas atom velocity and c is the speed of sound in the solid. For He/LiF Goodman shows that this ratio is about 0.15. Under these conditions it is a reasonable approximation to neglect the phonon tangential momentum compared with that of the gas atom. Furthermore for systems in which the periodicity of the surface is not important (e.g., metals) a continuum model of the solid may be appropriate, for which the only reciprocal lattice vector is $\underline{G} = \underline{0}$. It then follows from Eqs. (3.113) and (3.115) that

$$\underline{K}_r \simeq \underline{K}_i \tag{3.116}$$

which is a statement of the approximate conservation of tangential momentum of the scattering gas atom, as is assumed in the cube models.

The justification for assumption 1(a) in the soft-cube model comes from the fact that the attractive potential is due to a long-range interaction (primarily due to induced dipoles), and at any given position outside the solid probably has a significant contribution from several solid atoms. Assuming that the motions of the solid atoms are uncorrelated, we would expect the net interaction to be reasonably stationary with respect to the center of mass of the solid. Since we assume the potential to be stationary, its shape does not affect the interaction — for convenience, therefore, we may think in terms of a square-well potential. This potential serves to increase the velocity of the gas atom before the repulsive collision and decreases it again afterwards. Assumption 1(b) in the soft-cube model corresponds to the assumption widely used for gas-gas collisions that the repulsive potential has an exponential character (e.g., the Morse potential or Buckingham 6-exp potential). The exponential repulsion is clearly a better approximation than the impulsive potential used in the hard–cube model.

In assumption 1 it is also implied that the gas atom interacts with only a single surface atom. Since the impact points of the incident gas atoms tend to be distributed randomly over any unit cell of the surface, however, this may seem a questionable assumption. In Ref. [87] it is shown from consideration of the equipotential contours that for impact points over a large fraction of a unit cell, and hence for most collisions, most of the load is taken by a single surface atom. We also expect the interaction to affect more than one surface atom for grazing incident angles when the tangential velocity may be so high that the tangential distance travelled during the collision time is greater than a lattice spacing. An approximate calculation in Ref. [87] indicates that for the system Ar/Au with $\theta_i = 50°$ and $T_g = 2550°$K (data from Ref. [99]) the tangential distance covered during the repulsive collision is about half a lattice spacing. For a wide range of conditions the distance would be of this magnitude or less.

By means of assumption 3 for the soft–cube model we introduce into the analysis the important parameter ωt_c, where ω is the natural frequency of the surface atom, and t_c is a measure of the collision time. In the hard–cube model this parameter was effectively set equal to zero, and since the magnitude of the energy exchange is sensitive to the value of ωt_c, this was probably a major shortcoming of the hard–cube model. As discussed in Ref. [87] compared with the experimental results the hard–cube model tends to exaggerate the deviation from the specular of the maximum of the scattering pattern. These results imply that the hard–cube model tends to exaggerate the magnitude of the net energy exchange with the surface. Thus the use of the softer potential in the soft–cube model should improve the situation. Lattice interactions are still not taken into account in the model, however. Omission of lattice interactions can only really be justified for high–energy collisions where $\omega t_c \ll 1$, whereas for the thermal scattering regime it is found that typically $\omega t_c \simeq 1$. There is some evidence [100], however, from a comparison between a computer simulation of the interaction with a lattice of coupled oscillators and that for a lattice of independent oscillators (corresponding to our situation here) that the error in the energy change of the gas atom by neglecting the lattice is small ($\lesssim 10$ percent) in nearly all cases.

The mass ratio parameter $\mu = m_g/m_s$ gives an indication as to whether single collisions or multiple collisions of gas atoms with surface atoms are important. For light gas atoms striking a heavy surface ($\mu \ll 1$) a single collision is likely. For $\mu \gtrsim 1$ the single-collision models fail because many collisions with light surface atoms are needed to reverse the momentum of a heavy gas atom normal to the surface. In Ref. [85] it is pointed out that for the hard-cube model only single collisions with the surface can occur if $\mu < \frac{1}{3}$. A detailed discussion of the conditions for double collisions (up to $\mu = 1$) is given by Goodman [88] using the one-dimensional box model which formed the original basis of the hard-cube model. The analysis of the hard-cube model in Section 3.4.1. ignores multiple collisions and is therefore at best valid for $\mu < \frac{1}{3}$.

For the soft-cube model restrictions on the mass ratio are not so clear cut as for the hard-cube model because in addition to the possibility of multiple collisions, which is complicated by the presence of the spring, we have to consider the question of the validity of the perturbation procedure used. The collision of the gas atom with the surface atom in the soft-cube model is equivalent to the linear collision of a particle with a diatomic molecule (oscillator), which has been widely studied [101-105] in the field of chemical kinetics. A review of the theory of vibrational energy transfer between simple molecules in nonreactive collisions is given by Rapp and Kassal [47]. It is worth emphasizing that even in the case where the (harmonic) oscillator is initially stationary an exact closed-form analytical solution for the energy transfer cannot be obtained (it is basically a three-body problem), and hence one must either use a perturbation procedure, or else resort to numerical integration or iteration procedures [47]. We are here interested in comparing the perturbation procedure used for the soft-cube model in Section 3.4.2 with other perturbation procedures and with the exact solution. For simplicity we consider only the case where the surface is initially stationary, and compare the results for the energy transfer to the oscillator. We denote by ΔE_{sc} the energy transfer given by the perturbation procedure used for the soft-cube model. For the alternative perturbation procedure in which the surface atom is rigidly fixed for obtaining the zero-order trajectory we denote the energy transfer by ΔE_0; this corresponds more closely to the procedure usually used in the vibrational energy transfer problem [47]. ΔE_{exact} denotes the exact energy transfer. If we find γ directly from the perturbation procedure ($\gamma = \pi\omega b/u$), rather than the procedure used in Section 3.4.2, then it follows that

$$\Delta E_{sc} = (1 + \mu)^{-2}\Delta E_0 \tag{3.117}$$

It has been shown [47], however, that for $\mu \gtrsim 0.5$ the following approximate empirical relation holds:

$$\Delta E_{exact} = e^{-1.685\mu}\Delta E_0 \tag{3.118}$$

The interesting point is that for $\mu \gtrsim 0.5$ the expressions $(1 + \mu)^{-2}$ and $\exp(-1.685\mu)$ are approximately equal (within about 5% or better). Thus the basic perturbation procedure used in the soft-cube model gives good agreement with the exact calculation for an initially stationary oscillator for $\mu \gtrsim 0.5$. It is difficult to say anything specific for the case where the oscillator is initially excited. In the analysis in Section 3.4.2. a self-consistent procedure is used for obtaining γ. The above comparisons suggest that the value of γ obtained directly from the zero-order trajectory may be quite adequate, allowing a somewhat simpler analysis of the soft-cube model.

3.5. COMPUTER TRAJECTORY CALCULATIONS

In the structure scattering regime the incident energy of the gas atom is sufficiently high that the distance of closest approach of the center of the gas atom to that of the surface atom is small and the effective interaction surface is relatively rough. (See discussion in Section 3.4.4.) In this regime the surface temperature is unimportant as far as energy transfer is concerned, but as noted below (Section 3.5.2) it can influence the scattering through its effect on the instantaneous positions of the surface atoms. In the structure scattering regime the cube models, as discussed in Section 3.4, are not applicable. Virtually all the theoretical investigations in the structure regime use classical mechanics and rely on computer calculations of a large sample

of incident trajectories to obtain the scattering patterns. In the case of very
high incident energy of the gas atom a model in which both the gas atom and
the surface atoms are represented by free particles (hard spheres) is applicable. (We are not concerned here with very high incident energies,
$\simeq 100$ eV, for which sputtering and penetration of the solid may occur). We
consider the hard-spheres model in Section 3.5.1. At somewhat lower energies more general models with the surface atoms on springs and a realistic gas-solid interaction are more appropriate, as discussed in Section 3.5.2.

3.5.1. The Hard-Spheres Model

The hard-spheres model for structure scattering has been studied prin-
cipally by Goodman [106], Jackson and French [107], and most recently by
Busby et al. [108]. These treatments are basically similar, although the
work of Jackson and French is more general in that an attractive gas-surface potential well D is incorporated. Apart from that the gas-surface collision is calculated using purely hard-sphere mechanics. The surface atoms
are regarded as free particles, and they completely shield the solid atoms
below the surface layer. Because of the simple collision mechanics a greater
number of gas atom trajectories (each trajectory having a different aiming
point on the basic surface array) can be calculated with this model than with
other models for structure scattering. Goodman [106] has pointed out that
as many as 22,500 trajectories may be insufficient to yield a smooth scattering pattern.

A example of theoretical results from the hard-spheres model [107] for
the system Ar/W is shown in Fig. 3.16. Experimental points from O'Keefe
and French [109] are shown for comparison. The results in Fig. 3.16 are
for three different incident energies of the gas atom from 0.25 eV to 1.35 eV.
These energies are below the energy at which the hard-spheres model should
be strictly valid, but it is seen that there is quite good agreement between
theory and experiment. An important feature here is that the absolute value
of the flux is being compared. In most other theories (e.g., Section 3.4)
the scattering patterns are normalized so that only the shape is being compared, which is a less rigorous test. In making calculations with the

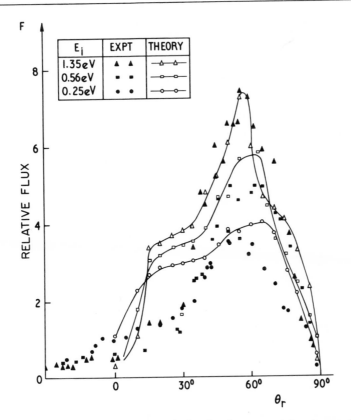

FIG. 3.16. Comparison between the hard-spheres model and experimental data. Experimental flux measurements from Ref. [109] for Ar/W(001), using monoenergetic incident beam (E_i values indicated), $\theta_i = 45°$, $T_s = 2000°$ K. A stationary attractive potential well, $D = 1.56$ kcal/mole, is used in the calculations, together with an effective hard-spheres radius of 4.1 Å. The calculated values are absolute values corresponding to the experimental geometry.

hard-spheres model a value must be chosen for the effective hard-spheres radius R (i.e., the distance of closest approach of the center of the gas atom to that of the surface atom). In Fig. 3.16 a value R = 4.1 Å is used. Other details are given in the figure caption.

3.5.2. More General Classical Models

These more general models have been studied by Oman and co-workers [110], Lorenzen, Raff, and McCoy [111] and McClure and Wu [112-113]. Although all these studies have contributed much to the understanding of scattering in this regime we here concentrate mainly on McClure's model [112-113], which is the most recent and probably best suited for producing scattering patterns of good angular resolution. As we see below the scattering patterns in this regime can have quite complicated multilobular forms and the question of statistical uncertainties in the results is important.

In McClure's model [112-113] the surface atoms in the region of the interaction are represented by a 2-dimensional array of independent oscillators. The gas atom interacts with each of these oscillators through a pairwise Lennard-Jones 6-12 potential

$$U = 4D[(s/r_j)^{12} - (s/r_j)^6]$$

$$(3.119)$$

where D is the pairwise binding energy, s is the range parameter and r_j is the distance from the gas atom to the jth surface atom. Outside this region of oscillators is a region of fixed force centers. The same Lennard-Jones potential applies for these fixed force centers. Thus as the gas atom approaches the surface it enters a region of 2-dimensional conservative potential, then passes into the region of interaction with the surface oscillators and then finally, if not trapped, passes again through the region of conservative potential as it leaves the solid. Criteria for determining adequate numbers of oscillators and fixed force centers are discussed in Ref. [112]. The force constant for the oscillator restoring potential in the tangential direction is obtained in terms of the bulk Debye temperature of the solid. The restoring force in the normal direction is taken as half the tangential value. The classical trajectory of the gas atom is calculated by integrating the equations of motion of the gas atom and the surface oscillators. Many gas atom trajectories must be calculated, with aiming points distributed over the surface. Non-zero surface temperature effects are taken into account by assigning a random initial state to each oscillator corresponding to the equilibrium energy distribution of an ensemble of oscillators. An example of the results of McClure's calculations are given in Section 3.5.3, in connection with rainbow scattering.

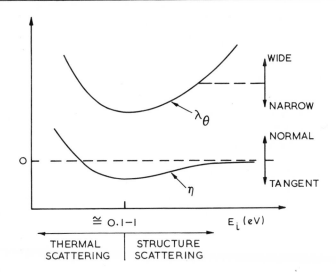

FIG. 3.17.   Qualitative behavior of λ_θ and η with E_i, illustrating transition from thermal to structure scattering (Goodman [114]).

The models of Oman [110], Lorenzen and Raff [111], and McClure [112-113] are sufficiently general to illustrate the transition from structure scattering to thermal scattering. In the thermal scattering regime, as typified for example by the results from the hard-cube model (Section 3.4.1), increasing incident gas energy is associated with the maximum of the scattering lobe moving away from the normal (and possibly below the specular) and the width of the lobe decreasing (as the spread due to thermal interaction with the surface becomes relatively less important). Thus, in terms of the deviation η of the maximum of the scattering lobe from the specular ($\eta \equiv \theta_i - \theta_{r\ max}$) and the half-width λ_θ of the lobe, we can say that $\partial\eta/\partial E_i$ and $\partial\lambda_\theta/\partial E_i$ are both negative in the thermal regime. In the structure regime, however, increasing E_i is associated with increased broadening of the lobe (as the structure seen by the gas atom increases) while the angular position of the maximum tends to a fixed value, at some position near the specular. At some intermediate E_i therefore we might expect to see a minimum in the values of both η and λ_θ. This is illustrated in Fig. 3.17 reproduced from Goodman's discussion [114] of this topic. Transition characteristics of this type are shown by the theoretical calculations we are considering. For example, results from McClure and

Wu [112] clearly show a minimum in λ_θ at an energy $E_i \simeq 1$ eV. Minima in
both λ_θ and η have also been observed experimentally by Miller and Subbarao
[14] (see also Ref. [114]), occurring at an energy $E_i \simeq 0.5$ eV. It is important
to note that thermal scattering and structure scattering represent two extreme forms of scattering behavior. Not all gas-surface systems will show a
clear cut transition from one form to the other. Systems with strongly periodic gas-surface potentials, such as some gas-ionic crystal systems, exhibit
structure scattering even at low values of E_i ($E_i \gtrsim 0.1$ eV) as shown in Fig.
3.19. Structure scattering here refers to effects arising from classical mechanics, as opposed to wave mechanical diffraction effects. For systems which
exhibit structure scattering at low E_i the surface temperature may be important through its disordering effect on the surface structure [112]. The situation is analogous to that which leads to the Debye-Waller factor in the quantum mechanical diffraction theory (see for example Ref. [55]).

3.5.3. Rainbow Scattering

The most important single feature to emerge from the computer trajectory calculations is the identification of the phenomena of surface rainbows.
Although the characteristic double lobes of rainbow scattering were observed
in several of the calculated scattering patterns [110-111], it was McClure
and Wu [112] who first suggested the correct interpretation of this phenomena.
Surface rainbows arise from classical mechanical scattering from a continuous periodic potential, as opposed to diffraction which is a quantum mechanical result of the periodicity. To explain the physical principle behind rainbow
scattering we refer to Fig. 3.18, based on a figure in Ref. [113]. For simplicity this shows a 2-dimensional model (with 1-dimensional periodicity), although similar arguments apply for the 3-dimensional model. The equipotential contour in Fig. 3.18(a) corresponds to a purely repulsive field. The distribution of incident trajectories crossing the x axis parallel to the surface
(away from the interaction region) is uniform. Thus the flux of atoms crossing an element of length dx is ndx, where n is a constant. Figs. 3.18(a) and
3.18(b) show the qualitative variation of θ_r with x. We see that over a distance x equal to a lattice spacing there are two turning points in θ_r as a function of x. The scattering distribution, i.e., the number of atoms scattered
into the angular range θ_r to $\theta_r + d\theta_r$, is given by $F(\theta_r)d\theta_r = n |dx/d\theta_r| d\theta_r$.

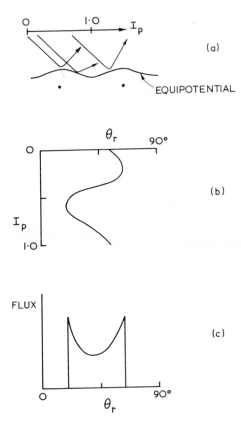

FIG. 3.18. Illustration of the physical principle of rainbow scattering (2-dimensional case), based on figure given by McClure [113]. I_p denotes the impact parameter, in units of a lattice constant.

It follows that $F(\theta_r) \to \infty$ at the angles θ_{r1} and θ_{r2} at which $d\theta_r/dx = 0$, giving rise to the rainbow peaks at these angles. From the above arguments it is seen that the mechanism of formation of surface rainbows applies generally for any potential of a continuous periodic nature. For the full 3-dimensional model out-of-plane rainbows also occur [113].

Scattering patterns exhibiting effects suggestive of surface rainbows have been observed experimentally, particularly for the rare gases on LiF [15, 22].

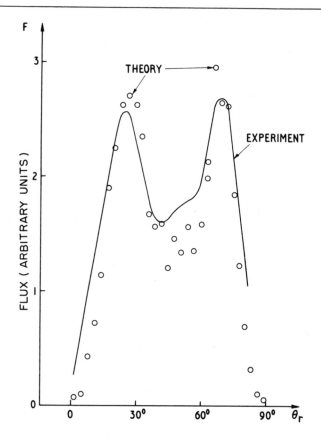

FIG. 3.19. Example of rainbow scattering. Comparison between McClure's model and experimental data for Ne/LiF(001) from Ref. [115]. $\theta_i = 51°$, $T_g = 940°$K (thermal beam) and $T_s = 300°$K. From Ref. [114].

Extensive comparisons between McClure's calculations [113, 115] and these experimental results have confirmed that the characteristic scattering features observed are caused by the surface rainbow mechanism discussed in the preceding paragraph. An example of a comparison between McClure's calculations and experiment is shown in Fig. 3.19 for Ne/LiF(001). The experimental conditions and values used for the calculation are shown in the figure caption. It is seen that the calculated points give the correct position

of the rainbow peaks to within a few degrees. Note also that here again absolute values of the flux are being compared, which makes the good agreement
even more significant.

3.6. PARAMETRIC MODELS AND
THE RECIPROCITY PRINCIPLE

Most of the theoretical models described in the previous sections contain
some free parameters, i.e., quantities involved in the interaction process
whose value is not known a priori, but is usually obtained by comparing with
experimental data. The parameters which occur in nearly all the models (except the hard-cube model) are the range and well-depth of the interaction potential. There is, however, a class of models in which additional and important details of the physical scattering process are described in terms of free
parameters. A simple and historically important example of this procedure
is Maxwell's assumption [116] that a fraction c of the incident atoms are
scattered diffusely and the remainder (1 - c) are scattered specularly. The
free parameter here is c. The advantage of this procedure is that the consequences of the scattering process in a wide range of conditions can conveniently be calculated mathematically, and then compared with experiment. In
Section 3.6.1 we describe some more recent and more complicated parametric models of the gas-surface scattering behavior. In Section 3.6.2 we discuss briefly the so-called principle of reciprocity, which has recently emerged
in connection with studies of gas-surface interactions. Although it is convenient to introduce this principle here in connection with the parametric models,
it is a general principle which has important implications for all gas-surface
scattering theories.

3.6.1. Parametric Models of Scattering Behavior

We consider first the model of Schamberg [117]. The scattering distribution is represented by a symmetric (conical) lobular distribution of angular

half-width λ_θ, centered about an angle θ_{r0}. The shape of the lobe, i.e., the flux of atoms scattered into the range θ_r to $\theta_r + d\theta_r$ is taken as

$$F(\theta_r)d\theta_r = K \cos\left(\frac{\theta_r - \theta_{r0}}{\lambda_\theta} \frac{\pi}{2}\right) \qquad (3.120)$$

The incident velocity of the gas atoms is u_i, and it is assumed that the velocity u_r of the scattered atoms is a constant, independent of θ_r. Considerations of flux conservation enable an expression for K to be obtained:

$$K = \frac{1}{2\pi} n_i u_i \sin\theta_i \left[\frac{1 - (\pi/2\lambda_\theta)^2}{1 - (\pi/2\lambda_\theta)^2 \sin\lambda_\theta} \right] \qquad (3.121)$$

The angle θ_{r0} is connected with the incident angle θ_i by

$$\cos\theta_{r0} = (\cos\theta_i)^\nu \qquad (3.122)$$

where $\nu \geq 1$, and u_r and u_i are connected by

$$u_r/u_i = (1 - \alpha)^{1/2} \qquad (3.123)$$

This model therefore contains the three parameters, ν, λ_θ, and α. With this model Schamberg made extensive calculations on the free molecule aerodynamics of cones, spheres, flat plates, and other shapes. This work has been used in calculating limits on drag coefficients of orbiting bodies, e.g., in Ref. [118].

A more sophisticated parametric model, similar in some respects to that of Schamberg, has been proposed by Goodman [119]. Goodman uses the results from his hard-spheres model [106] (section 3.5.1) to put the parametric model on a firmer physical basis. For example the results from the hard-spheres model are used (empirically) to provide an expression for the speed distribution of the scattered atoms, giving a more physically realistic result than Schamberg's assumption that the outgoing velocity u_r is constant for all outgoing angles. Similarly Goodman obtains probability distributions for the in-plane and azimuthal scattering angles. The full development of the model is quite involved, and the reader is referred to the original paper [119] for full details. To the author's knowledge, no aerodynamic calculations using this model have been published.

Another parametric model for gas–surface scattering has been proposed by Nocilla [120]. Nocilla assumes that the velocity distribution of the re-emitted atoms, i.e., atoms scattered from the surface, is given by the following displaced Maxwellian distribution

$$f_+(\underline{q}) = \frac{n_r}{(2\pi k T_r)^{3/2}} \exp \frac{-(\underline{q} - \underline{U}_r)^2}{2kT_r} \tag{3.124}$$

where $\underline{q}$ is the molecular velocity, $\underline{U}_r$ is the macroscopic flow velocity of the outgoing stream, and T_r is the effective temperature of the distribution. The normalization constant n_r is found from considerations of flux conservation. It is convenient to define a speed ratio $S_r = U_r/\sqrt{2kT_r}$ and an angle θ_r^*, where θ_r^* is the direction of the vector $\underline{U}_r$. The distribution of scattered atoms is then fully specified in terms of the three parameters T_r, S_r, and θ_r^*. The normalized scattering distributions are in fact described by the two parameters S_r and θ_r^*. By choosing values for these two parameters Nocilla [120] has matched this model with the lobular scattering patterns for the system Ar/LiF; very good agreement with the shape of the scattering patterns can be obtained. Comparison with velocity or energy transfer measurements would be required for an evaluation of the parameter T_r. It is clear that the Nocilla model takes no real account of the physical details of the gas–surface collision. One major drawback of the model is that there is no correlation between the distributions of the incident and reflected molecules. Despite this the model has been used [121] for calculations of the lift and drag of bodies in free molecular flow.

Another model proposed by Nocilla and Chiado-Piat [122] is closer in spirit to the hard-cube model (Section 3.4.1) than to that described in the preceding paragraph. As in the original treatment of the hard-cube model [85] the surface atoms are assumed to have a velocity $v_i = kT_s/m_s\bar{u}_{ni}$ outward along the surface normal. It is assumed, however, that the motion of the surface atom is unchanged during the collision, i.e., it has infinite mass as far as the collision is concerned. The outer surface of the cubes is assumed to be randomly sloped (similar to the situation used in Ref. [86]) to simulate surface roughness on an atomic scale. The major difference from the hard-cube model, and the main justification for classifying this model as

a parametric model, is that the outgoing velocity of the gas atoms calculated
on the basis of an impulsive collision is reduced by a factor h, which is introduced to account for the softness of the interaction potential and the lattice effects. Using this model expressions are obtained [122] for the velocity distribution of the scattered atoms, and the momentum and energy accommodation coefficients. Comparison between theory and experiment yields
values of the dissipation parameter h. In order to obtain good agreement is
is necessary to allow h to be a function of T_s, which indicates that such a
simple parametric description cannot fully account for the complicated physical details of the collision.

3.6.2. Principle of Reciprocity in Gas-Surface Scattering

The principle of reciprocity has emerged [123-128] from studies of the
boundary conditions for the Boltzmann equation in problems of rarefied gas
dynamics. These boundary conditions describe the interaction of the gas
molecules with the solid walls bounding the region where the gas flows. An
understanding of the scattering, or reemission, behavior of the gas molecules is essential for setting up these boundary conditions. As indicated in
Section 3.6.1 the parametric models have been used for this purpose.

Full derivations of the reciprocity relations are given by Kuščer [124]
and Wenaas [128]. We give here only a brief discussion of the basic principles and the implications of the result. It is convenient here to adopt a
notation for the velocity of the gas atoms somewhat different from that used
elsewhere in this article, in order to be more consistent with the usual con-
ventions of kinetic theory. Thus we use $\underline{q}'$ and $\underline{q}$ to denote the velocity vectors of the gas atoms before and after collision with the surface, respectively.
The components of these vectors along the z axis (the outward normal) are
q_z' and q_z. The velocity distribution f_+ of the gas atoms scattered from the
surface can then be expressed in terms of the incident distribution f_- by the
equation

$$\underline{q}_z f_+(\underline{q}) = \int_- |q_z'| f_-(\underline{q}') P(\underline{q}' \to \underline{q}) d^3 q' \qquad (3.125)$$

where $P(\underline{q}' \to \underline{q})$ represents the probability for an atom of initial velocity $\underline{q}'$ to be scattered into d^3q near $\underline{q}$. This probability kernel is only defined for $q_z' < 0$, and this restriction is signified by the minus sign on the integral. Equation (3.125) applies if f_+ and f_- are time-independent: the more general time-dependent case is discussed in Ref. [124]. The linear relationship between f_- and f_+, as implied by Eq. (3.125) should be valid so long as the collision time of the atom is short enough and the gas sufficiently rarefied, so that the atoms are scattered independently of one another. Some discussion of the nonlinear situation is given in Ref. [124]. It is assumed that every atom is reflected, so that P is normalized:

$$\int_+ P(\underline{q}' \to \underline{q})d^3q = 1 \qquad (3.126)$$

Irrespective of the scattering mechanism the kernel P obeys the following reciprocity relation, or detailed balance relation:

$$|q_z'| M(\underline{q}')P(\underline{q}' \to \underline{q}) = q_z M(\underline{q})P(-\underline{q} \to -\underline{q}') \qquad (3.127)$$

where

$$M(\underline{q}) = \left(\frac{m_g}{2\pi k T_s}\right)^{3/2} \exp \frac{-m_g q^2}{2k T_s} \qquad (3.128)$$

is the Maxwellian distribution with the surface temperature. Reciprocity follows in a rigorous and completely general way from time-reversal invariance, and from the assumption of thermal equilibrium for the solid. Both classical and quantum mechanical derivations are given by Kuščer [124] and Wenaas [128].

The detailed balance condition has already been used in the quantum mechanical gas-surface collision theories for the transition probabilities to a single oscillator [Eq. (3.13)]. Equation (3.127) is more general, however, being applicable for any multiphonon process. Reciprocity, together with normalization and nonnegativity of the scattering kernel, imposes severe restrictions on any parametric models used for a description of the gas-surface interaction. For example, Nocilla's model based on a displaced Maxwellian distribution (Section 3.6.1.) is only consistent with reciprocity for small deviations from equilibrium with the surface. Further parametric models, taking the reciprocity restriction into account, are under development [125-127]. Recently Miller and Subbarao [129] have given a direct experimental verification of the reciprocity principle, using the scattering of Ne and Ar beams from an Ag(111) surface. By making density and velocity measurements of the scattered beam at various angles they have confirmed that (within the experimental error) the reciprocity relation Eq. (3.127) is obeyed.

4. TRAPPING EFFECTS IN GAS-SURFACE COLLISIONS

In the preceding sections we discussed the various gas-surface collision theories in relation to the scattering distributions and the energy accommodation coefficient, without much discussion of trapping (or sticking, or condensation) probabilities (or coefficients). This is mainly because the more comprehensive and reliable experimental data available are for the scattering distributions and accommodation coefficient. There are, however, some experimental data available on trapping probabilities and, as the ability to calculate this quantity is one of the goals of gas-surface collision theory, we give in this section a brief discussion of the current theoretical situation. In principle any of the theoretical approaches in Section 3 (except the parametric models) could form the basis of a calculation of the trapping probability. In practice it has proved difficult to make reliable calculations of the trapping probability, even for simple systems involving rare gases. There are several reasons for this situation, the main ones being (1) that we have only limited a priori quantitative knowledge about the gas-surface interaction potential, and the calculated trapping probability is very sensitive to the details of this potential (e.g., the well depth); and (2) that it is difficult to make an unambiguous definition of trapping which is open to both theoretical calculation and experimental measurement. In Section 4.1 we consider the question of a definition of trapping and discuss general physical features of trapping behavior. In Section 4.2 we consider several specific theoretical calculations of the trapping probability.

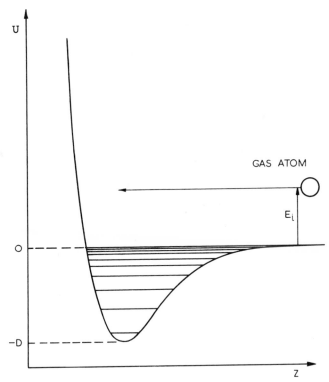

FIG. 4.1. Representative gas–surface interaction potential, as function of distance z normal to the surface.

4.1. GENERAL CONCEPTS OF TRAPPING

Various different definitions of trapping can be made, but probably the most satisfactory for the present discussion is the following simple one given by Goodman [130]: "trapping of a gas atom at a solid surface occurs if the (total) energy of the gas atom becomes negative." In this definition, a stationary gas atom at an infinite distance from the surface is taken as having zero energy (see Fig. 4.1). Once the gas atom is trapped according to this definition it cannot escape from the surface until it regains sufficient energy (and momentum normal to the surface) as a result of the continuing

"collisions" with the surface as the atom oscillates in the potential well. It
is through these collisions within the potential well that the trapped (adsorbed)
atoms tend to come into equilibrium with the solid.

As mentioned in Section 3.4.4, and discussed by Goodman [68], the long-
range attractive part of the interaction is to a good approximation conservative, so that the gas atom exchanges energy with the solid only during the
short-range repulsive part of the interaction, beginning at the bottom of the
potential well. In the present discussion the term "collision" refers to this
interaction with the repulsive part of the potential. In many calculations it
is assumed that the gas atom is trapped if it experiences more than one collision with the surface. Even for the case of a cold solid ($T_s \ll T_g$) this may
not be a good assumption, however, from the point of view of comparing with
experimental results, because it is quite possible for the gas atom to gain
momentum tangential to the surface at the expense of its momentum normal
to the surface, while its total energy remains positive. It may continue in
this state for several collisions until either its energy is reduced to a negative value or else its tangential momentum is transformed back into normal
momentum, sufficient to escape from the surface. The motion of the gas
atom during these collisions has been called "hopping" by Goodman [130].
This motion can only be analyzed with models which follow the motion of the
gas atoms in more than one dimension. The hopping motion is very difficult
to calculate, but an attempt in this direction has been made by Goodman [130],
as discussed in Section 4.2.

A related problem which has recently been studied [131] is the nonequilibrium behavior of an ensemble of gas atoms adsorbed on a solid surface.
Under conditions where the adsorbed atoms are in thermal equilibrium with
the solid (and the external gas phase) the population of adatoms in the energy
levels of the potential well (Fig. 4.1) would be given by a time-independent
distribution, dominated by a Boltzmann factor $\exp(-E/kT_s)$. Under nonequilibrium conditions the distribution may be quite different. For example,
in a steady-state adsorption (condensation) process the upper energy levels
in the potential well will be relatively highly populated compared with the
equilibrium distribution. The extent to which the population deviates from
the equilibrium distribution depends to a large extent on the efficiency of the

repulsive collision in exchanging energy between the adsorbed atom and the solid. Details of the calculations and some preliminary results are given in Ref. [131]. Obviously, the hopping atoms of the preceding paragraph are closely related to atoms in (nonequilibrium) upper levels of the potential well.

4.2. SPECIFIC TRAPPING CALCULATIONS

The first detailed calculations [61-63] of the trapping probability in a gas-surface collision were based on a one-dimensional semi-infinite lattice model of the solid. The procedure for solving the equations of motion has already been discussed in Section 3.2. In the studies in Refs. [61]-[63] simple interaction potentials are used (mostly the truncated harmonic oscillator potential) for which the equations of motion can be solved exactly. Even for the truncated harmonic potential mathematically convenient solutions only occur for mass ratios $\mu = \frac{1}{2}$ and $\mu = 1$. In Ref. [63] numerical results are given for other mass ratios. In all these calculations the lattice is initially cold ($T_S = 0$); if the incident energy of the gas atom (associated with motion normal to the surface) is less than some critical value E_C it has insufficient energy after the collision to return to the cut-off separation of the harmonic potential, and hence is trapped. For energies greater than E_C the gas atom always escapes from the solid after the collision, but with somewhat reduced energy. Numerical results from Zwanzig [62], which are borne out in the other papers [61, 63], are

$$\mu = \frac{1}{2}; \quad E_c/D \simeq 2.4$$

$$\mu = 1; \quad E_c/D \simeq 24.5$$

where D is the well-depth.

The result for $\mu = 1$ is often quoted in connection with condensation processes in the form "the gas atom will lose sufficient energy to the lattice to be captured if its incident energy is equal to or less than 25 times the heat of adsorption." In view of the inadequacy of the one-dimensional lattice model (see Section 3.2.2) and the unrealistic interaction potential used, however,

this statement cannot really be trusted. As discussed in Section 3.2.2 the
one-dimensional lattice tends to allow too much energy to be transferred to
the solid, thus enhancing the trapping probability. This trend is confirmed
in Ref. [63], where a comparison with experiment for the system Xe/W shows
that the one-dimensional model gives too high a value for the trapping probability.

A more recent calculation of the critical energy for trapping E_c has been
made by Goodman [68], using a three-dimensional lattice model with a realistic (Morse) interaction potential (as described in Section 3.2.1). The calculations use a numerical iteration procedure and so should be applicable for
large mass ratios and for the large energy transfers involved in trapping.
As discussed in Section 3.2.2 the primary object of these calculations is to
compare with the experimental data on the energy accommodation coefficient.
In the process, however, one must calculate the fraction of gas atoms which
are trapped (it is assumed that these are completely accommodated to the
surface temperature $T_s = 0$). The values for the range and well-depth parameters in the potential are obtained by the matching with the accommodation coefficient data. Values for the fraction of gas atoms trapped at $T_g =$
$300°$ K ($T_s = 0$) from Goodman's [68] calculations for the systems He, Ne,
Ar, Kr and Xe/W are 0.00, 0.01, 0.20, 0.38, and 0.70 respectively.
Covington and Ehrlich [132] have experimentally determined a value of
approximately 0.4 for the sticking coefficient of Xe at $300°$ K on a tungsten
surface at $80°$ K.

A gas atom of given energy colliding with the surface of a solid at $0°$ K,
assuming an idealized head-on collision as in the lattice models described
in the preceding paragraphs, has a probability of either 0 or 1 of being
trapped, depending on whether its incident energy (associated with motion
normal to the surface) is less than or greater than some critical energy E_c.
The concept of a fraction of atoms trapped arises through consideration of
the distribution of energies in the incident gas. We now consider the situation where a gas atom of fixed energy collides with the surface of a warm
solid ($T_s > 0$). The trapping probability for the gas atom is no longer exactly 0 or 1 in general, and varies with the incident energy E_i and the temperature T_s of the solid. Probably the most suitable model with which to

study this situation is the soft-cube model [80, 87] (Section 3.4.2). The hard-cube model has also been used [133] for trapping calculations, but this cannot be expected to represent the energy transfer process as accurately as the soft-cube model. A trapping calculation for a one-dimensional lattice with $T_s > 0$ has been made by Armand [76], using a truncated harmonic potential as in the early studies described above. According to the soft-cube model the probability that a gas atom with incident normal component of velocity u_{ni} will have insufficient energy to leave the surface after its first collision (i.e., will be "initially trapped") is given by

$$S_i = \frac{1}{2} \left\{ 1 - \text{erf} \left[\left(\frac{m_s u_{ni}'^2 v_{c0}^2}{2kT_s} \right)^{1/2} \right] \right. $$

$$ \left. - \frac{1}{\sqrt{\pi}} \left(\frac{2kT_s}{m_s u_{ni}'^2} \right) \exp \left(- \frac{m_s u_{ni}'^2 v_{c0}^2}{2kT_s} \right) \right\} \tag{4.1}$$

where u_{ni}' is the normal component of velocity within the potential well,

$$u_{ni}'^2 = u_{ni}^2 + \frac{2D}{m_g} \tag{4.2}$$

and v_{c0}, the initial velocity of the oscillator corresponding to an outgoing velocity of the gas atom of zero, is [from Eq. (3.102)]

$$v_{c0} \equiv \frac{v \cos \omega t_0}{u_{ni}'} = \frac{1}{2J} \left[\left(\frac{u_{ni}}{u_{ni}'} \right)^2 - \mu J^2 \right] \tag{4.3}$$

where J is given by Eq. (3.87). The result in Eq. (4.1) follows from Eq. (3.103) by noting that

$$S_i \equiv 1 - W(0, \mathscr{E}_i)$$

The soft-cube model does not include hopping effects, so that S_i is not necessarily equal to the trapping probability defined in Section 4.1. To make numerical calculations from Eq. (4.1) we need values from the range and well depth of the interaction potential. Some calculated values of S_i are shown in Table 4.1, taken from Ref. [87], where the required values of the potential parameters were obtained by matching with experimental scattering data. There are no experimental data with which to directly compare

TABLE 4.1

Values of the Initial Trapping Probability S_i Calculated from the Soft-Cube
Model, Using n and D Values As Given in Table 3.3

System	T_g, °K	T_s, °K	S_i
He/Au	300	600	0.00
	2500	600	0.00
Ne/Au	300	600	0.02
	2550	600	0.00
Ar/Au	300	600	0.15
	2550	600	0.00
Xe/Au	300	600	0.66
	2550	600	0.25
Ar/Ag	300	560	0.11
	1550	560	0.01
Xe/Ag	300	560	0.40
	1500	560	0.17

these results. The results for Xe/Au and Xe/Ag are roughly in line with
those of Goodman [68] and Covington and Ehrlich [132] mentioned above.
Some further trapping calculations based on the soft-cube model are given
in Ref. [80] (see Figs. 3.12 and 3.13).

Goodman [130] has made a quantitative estimate of the effect of the hopping motion described in Section 4.1, using his empirical velocity distribution function of the scattered molecules (Section 3.6.1) which is based on his
hard-spheres model (Section 3.5.1). The gas atom is allowed to have as
many collisions with the repulsive potential as necessary until it either escapes from the surface or is trapped. Rather than follow the actual trajectory of the gas atom through this series of collisions, the conditions before
each collision are chosen in a random manner from a specified distribution.

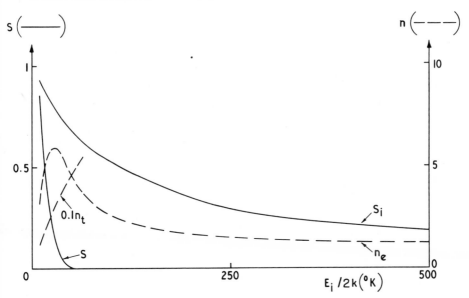

FIG. 4.2. Calculated values of S, S_i, n_t, and n_e versus $E_i/2k$ for the system Ne/Ag, using a value $D/k = 146°$ K. The results are for monoenergetic incident beams with $\theta_i = 50°$ and $T_s = 0°$ K (Goodman [130]).

Some results from Goodman's calculations are shown in Fig. 4.2, for the system Ne/Ag with $T_s = 0$; the results are for an incident monoenergetic beam ($E_i \equiv 2kT_i$) with $\theta_i = 50°$. The quantities shown in the figure are: S, the true trapping probability (according to the definition in Section 4.1), obtained by allowing each gas atom to undergo as many hops as necessary until it is either trapped or escapes; S_i, the probability of undergoing more than one collision with the surface (i.e., the probability of being "initially trapped"); n_t, the average number of collisions made by a gas atom which is eventually trapped; n_e, the average number of collisions made by a gas atom which eventually escapes. The important feature in Fig. 4.2 is not so much the absolute values, but the large difference between S and S_i.

Little progress has been made in calculating trapping probabilities using the quantum mechanical gas–surface collision theories (Section 3.1). There are several reasons for this state of affairs. Until recently the quantum

theories have relied on a first-order perturbation treatment, which is probably not really suitable for transitions from a free to a bound state. Furthermore most of the theories until recently have allowed only single-phonon
transitions and, as pointed out by Zwanzig [62] and others, many condensation processes (particularly from a high temperature) must involve creation of several phonons (using the Debye temperature for an estimate of the
maximum energy of a single phonon). Another relevant point is that the
theoretical treatment of Lennard-Jones and Devonshire (Section 3.1) only
includes transitions from free incident to free outgoing states of the gas
atom. Only recently has this basic theory been extended [48] to include
transitions involving bound states. Quantum mechanical calculations of
trapping probabilities, particularly for the rare gases at low temperatures,
would be worthwhile.

LIST OF SYMBOLS

The symbols which appear most frequently are listed below. Some other
symbols are defined in the text for local use.

a	lattice spacing
a_0	Bohr radius
B	pre-exponential of repulsive potential
b	range parameter of repulsive potential
c_1, c_2	speeds of sound in solid
D	well depth of interaction potential
D*	dimensionless well depth $2D/m_g u_{ni}^2$
$\mathscr{D}$	dimensionless well depth D/kT_g
d	$2m_g D/\kappa^2 \hbar^2$, parameter of Morse potential
E	energy of gas atom
E_{cr}	critical energy for trapping
$\mathscr{E}$	dimensionless energy E/kT_g
F	force between gas atom and surface
$F(\theta_r)$	flux of atoms scattered in unit angular range at θ_r
f_-, f_+	velocity distribution functions of incident and outgoing gas atoms respectively

f_t	fraction of gas atoms trapped
$f_\theta(u)$	speed distribution of scattered atoms at angle θ
$\underline{G}$	reciprocal lattice vector
$G(\underline{\theta}, \tau)$	normal mode transform
H	Hamiltonian
H'	perturbing Hamiltonian
I	dimensionless impulse integral
J	energy transfer integral
$\underline{K}$	gas atom wave vector parallel to surface
$\underline{k}$	wave vector of gas atom
k_{ij}, k_0	lattice force constants
$\underline{L}_n$	tangential component of ℓ_n
$\underline{\ell}_n$	phonon wave vector
ℓ_0	characteristic length (for area of stress pulse)
m	atomic mass
m_r	reduced mass, $m_s m_g/(m_g + m_s)$
N	number of surface atoms
$\underline{N}$	position of atom in lattice
n	(b/a_0) $(\hbar\omega/k\,\Theta_D)$
P	transition probability
p	gas atom momentum, $2m_g E/\kappa^2\hbar^2$
$\underline{q}'$, $\underline{q}$	velocity vectors of gas atom before and after collision with surface, respectively
q_z', q_z	components along outward normal of $\underline{q}'$ and $\underline{q}$, respectively
$\underline{R}$	position of solid atom
$\underline{r}$	position of gas atom
r	gas–surface separation in normal direction, $z_g - z_s$
S	trapping probability
S_i	initial trapping probability
T	temperature
t	time
t_c	characteristic collision time
U	gas–surface interaction potential
u_i	incident velocity of gas atom
u_{ni}	normal component of u_i
u_{ti}	tangential component of u_i

u_r, u_{nr}, u_{tr} outgoing velocities corresponding to u_i, u_{ni}, u_{ti}, respectively

$u_{ni}{}'$, $u_{nr}{}'$ velocities within the potential well corresponding to u_{ni} and u_{nr}, respectively

V gas–surface interaction potential

v velocity of surface atom (normal to surface)

v_c $v \cos \omega t_0 / u_{ni}{}'$

v_s $v \sin \omega t_0 / u_{ni}{}'$

$W(\mathscr{E}_r, \mathscr{E}_i)$ probability of gas atom leaving surface with $\mathscr{E} > \mathscr{E}_r$

$w(n)$ probability of occupation number n in lattice mode

X response function of lattice

x coordinate in surface plane

y coordinate in surface plane

z coordinate normal to surface

z_{gf} double integral of force $F(t)$

α energy accommodation coefficient

α_{lim} α in the limit $T_g \to T_s$

γ soft–cube parameter ($= \pi \omega t_c / 2$)

δ $[D^*/(1 + D^*)]^{1/2}$

η difference between specular angle and the angular position of the maximum of the scattering pattern ($\theta_i - \theta_{r\ max}$)

Θ_D Debye temperature of solid

θ_i, θ_r incident and outgoing angles of gas atom (measured from surface normal in plane containing incident beam and the normal)

$\underline{\theta}$ normal mode coordinate

κ inverse range parameter of Morse potential

λ wavelength

λ_θ half–width of scattering distribution

μ mass ratio m_g / m_s

ν_D Debye frequency

ρ density of states

σ $\omega b / u_{ni}{}'$

τ dimensionless time $2(k_0/m_s)^{1/2} t$

τ^* dimensionless time t/t_c

ϕ_α eigenstate of average gas-surface potential

χ response function of 3D infinite lattice

ψ wave function

ω frequency of oscillator mode

Subscripts

f   final state

g   relating to gas atoms

i   incident condition of gas atom or initial state

q   mode of lattice

r   reflected (outgoing) condition of gas atom

s   relating to surface

REFERENCES

[1] L. I. Schiff, Quantum Mechanics (2nd ed.), McGraw-Hill, New York, 1955.

[2] J. O. Hirschfelder, C. F. Curtiss, and R. B. Bird, Molecular Theory of Gases and Liquids, Wiley, New York, 1964.

[3] D. M. Young and A. D. Crowell, Physical Adsorption of Gases, Butterworths, London, 1962.

[4] M. Kaminsky, in Atomic and Ionic Impact Phenomena on Metal Surfaces, Academic, New York, 1965.

[5] J. H. de Boer, Advan. Catal. 8, 18 (1956).

[6] E. C. Beder, Advan. Atomic Mol. Phys., 3, 205 (1967).

[7] D. M. Gilbey, in Rarefied Gas Dynamics, Proc. 5th Intern. Symp., Oxford, 1966 (C. L. Brundin, ed.), Vol. 1, Academic Press, New York, 1967, p. 121.

[8] F. O. Goodman, Surface Sci., 11, 283 (1968).

[9] S. A. Schaaf and P. L. Chambre, Flow of Rarefied Gases, Princeton Univ. Press, Princeton, New Jersey, 1961.

[10] S. Yamamoto and R. E. Stickney, J. Chem. Phys., 53, 1594 (1970).

[11] R. E. Stickney, Advan. Atomic Mol. Phys., 3, 143 (1967).

[12] J. N. Smith Jr., H. Saltsburg, and R. L. Palmer, J. Chem. Phys.,
     49, 1287 (1968).

[13] S. S. Fisher, O. F. Hagena, and R. G. Wilmoth, J. Chem. Phys., 49,
     1562 (1968).

[14] D. R. Miller and R. B. Subbarao, J. Chem. Phys., 52, 425 (1970).

[15] J. N. Smith Jr., D. R. O'Keefe, and R. L. Palmer, J. Chem. Phys.,
     52, 315 (1970).

[16] D. R. O'Keefe, J. N. Smith Jr., R. L. Palmer, and H. Saltsburg,
     J. Chem. Phys., 52, 4447 (1970).

[17] D. R. O'Keefe, J. N. Smith Jr., R. L. Palmer, and H. Saltsburg,
     Surface Sci., 20, 27 (1970).

[18] V. S. Calia and R. A. Oman, J. Chem. Phys., 52, 6184 (1970).

[19] R. L. Palmer, J. N. Smith Jr., H. Saltsburg, and D. R. O'Keefe,
     J. Chem. Phys., 53, 1666 (1970).

[20] D. L. Smith and R. P. Merrill, J. Chem. Phys., 53, 3588 (1970).

[21] M. J. Romney and J. B. Anderson, J. Chem. Phys., 51, 2490 (1969).

[22] J. N. Smith Jr., D. R. O'Keefe, H. Saltsburg, and R. L. Palmer,
     J. Chem. Phys., 50, 4667 (1969).

[23] L. B. Thomas, in Fundamentals of Gas-Surface Interactions,
     (H. Saltsburg, J. N. Smith Jr., and M. Rogers, eds.), Academic
     Press, New York, 1967, p. 346.

[24] H. Y. Wachman, J. Amer. Rocket Soc., 32, 2 (1962).

[25] M. Knudsen, The Kinetic Theory of Gases, Methuen, London (1950).

[26] L. B. Thomas and R. C. Golike, J. Chem. Phys., 22, 300 (1954).

[27] F. O. Goodman, Surface Sci., 24, 667 (1971).

[28] J. E. Lennard-Jones and A. F. Devonshire, Proc. Roy. Soc., Ser. A,
     156, 6 (1936).

[29] J. E. Lennard-Jones and A. F. Devonshire, Proc. Roy. Soc., Ser. A,
     156, 29 (1936).

[30] J. E. Lennard-Jones and A. F. Devonshire, Nature, 137, 1069 (1936).

[31] J. E. Lennard-Jones and A. F. Devonshire, Proc. Roy. Soc., Ser. A,
     158, 242 (1937).

[32] J. E. Lennard-Jones and A. F. Devonshire, Proc. Roy. Soc., Ser. A,
     158, 253 (1937).

[33] A. F. Devonshire, Proc. Roy. Soc., Ser. A, 156, 37 (1936).

[34] A. F. Devonshire, Proc. Roy. Soc., Ser. A, 158, 269 (1937).

[35] J. E. Lennard-Jones and C. Strachan, Proc. Roy. Soc., Ser. A., 150, 442 (1935).

[36] J. E. Lennard-Jones and C. Strachan, Proc. Roy. Soc., Ser. A, 150, 456 (1935).

[37] J. E. Lennard-Jones and C. Strachan, Proc. Roy. Soc., Ser. A, 158, 591 (1937).

[38] C. Zener, Phys. Rev., 37, 557 (1931).

[39] C. Zener, Proc. Roy. Soc., 40, 178 (1932).

[40] C. Zener, Proc. Roy. Soc., 40, 335 (1932).

[41] J. M. Jackson, Proc. Cambridge Phil. Soc., 28, 136 (1932).

[42] J. M. Jackson and N. F. Mott, Proc. Roy. Soc., Ser. A, 137, 703 (1932).

[43] J. M. Jackson and A. Howarth, Proc. Roy. Soc., Ser. A, 142, 447 (1933); 152, 515 (1935).

[44] D. M. Gilbey, J. Phys. Chem. Solids, 23, 1453 (1962).

[45] R. T. Allen and P. Feuer, in Rarefied Gas Dynamics, Proc. 5th Intern. Symp., Oxford, 1966 (C. L. Brundin, ed.), Vol. 1, Academic Press, New York, 1967, p. 109.

[46] B. Baule, Ann. Phys., 44, 145 (1914).

[47] D. Rapp and T. Kassal, Chem. Rev., 69, 61 (1969).

[48] F. O. Goodman and J. D. Gillerlain, in Rarefied Gas Dynamics, Proc. 7th Intern. Symp., Pisa, 1970, (to be published); and J. Chem. Phys., 54, 3077 (1971).

[49] R. Manson and V. Celli, Surface Sci., 24, 495 (1971).

[50] I. Estermann and O. Stern, Z. Phys., 61, 95 (1930).

[51] H. S. W. Massey and E. H. S. Burhop, Electronic and Ionic Impact Phenomena, Oxford Univ. Press, Oxford, England, 1952.

[52] R. Frisch and O. Stern, Z. Phys., 84, 430 (1933).

[53] E. C. Beder, Surface Sci., 1, 242 (1964).

[54] A. Tsuchida, Surface Sci., 14, 375 (1969).

[55] N. Cabrera, V. Celli, F. O. Goodman, and R. Manson, Surface Sci., 19, 67 (1970).

[56] F. O. Goodman, Surface Sci., 19, 93 (1970).

[57] M. Gell-Mann and M. L. Goldberger, Phys. Rev., 91, 398 (1953).

[58] T. Y. Wu and T. Ohmura, Quantum Theory of Scattering, Prentice-Hall, Englewood Cliffs, New Jersey, 1962.

[59] S. S. Fisher, M. N. Bishara, A. R. Kuhlthau, and J. E. Scott Jr., in Rarefied Gas Dynamics, Proc. 6th Intern. Symp., Cambridge, Mass., 1968 (L. Trilling and H. Wachman, eds.), Vol. 2, Academic Press, New York, 1969, p. 1227.

[60] D. R. O'Keefe, R. L. Palmer, H. Saltsburg, and J. N. Smith Jr., J. Chem. Phys., 49, 5194 (1968).

[61] N. Cabrera, Disc. Faraday Soc., 28, 16 (1959).

[62] R. W. Zwanzig, J. Chem. Phys., 32, 1173 (1960).

[63] B. McCarroll and G. Ehrlich, J. Chem. Phys., 38, 523 (1963).

[64] B. McCarroll, J. Chem. Phys., 39, 1317 (1963).

[65] F. O. Goodman, J. Phys. Chem. Solids, 23, 1269 (1962).

[66] F. O. Goodman, J. Phys. Chem. Solids, 23, 1491 (1962).

[67] F. O. Goodman, J. Phys. Chem. Solids, 24, 1451 (1963).

[68] F. O. Goodman, in Rarefied Gas Dynamics, Proc. 4th Intern. Symp., Toronto, 1964, (J. H. De Leeuw, ed.), Vol. 2, Academic Press, New York, 1966, p. 366.

[69] F. O. Goodman, Surface Sci., 3, 386 (1965).

[70] F. O. Goodman, Surface Sci., 5, 283 (1966).

[71] F. O. Goodman, Surface Sci., 11, 283 (1968).

[72] L. Trilling, J. Mec., 3, 215 (1964).

[73] L. Trilling, in Rarefied Gas Dynamics, Proc. 5th Intern. Symp., Oxford, 1966 (C. L. Brundin, ed.), Vol. 1, Academic Press, New York, 1967, p. 139.

[74] C. M. Chambers and E. T. Kinzer, Surface Sci., 4, 33 (1966).

[75] A. R. Ducharme and E. T. Kinzer, Surface Sci., 17, 69 (1969).

[76] G. Armand, Surface Sci., 9, 145 (1968); also in Rarefied Gas Dynamics, Proc. 6th Intern. Symp., Cambridge, Mass., 1968 (L. Trilling and H. Wachman, eds.), Vol. 2, Academic Press, New York, 1969, p. 1055.

[77] F. O. Goodman and H. Y. Wachman, J. Chem. Phys., 46, 2376 (1967).

[78] L. B. Thomas and E. B. Schofield, J. Chem. Phys., 23, 861 (1955).

[79] L. B. Thomas, in Rarefied Gas Dynamics, Proc. 5th Intern. Symp., Oxford, 1966 (C. L. Brundin, ed.), Vol. 1, Academic Press, New York, 1967, p. 155.

[80] R. M. Logan, Surface Sci., 15, 387 (1969).

[81] L. Trilling, Surface Sci., 21, 337 (1970).

[82] L. D. Landau, Phys. Z. Sowjetunion, 8, 489 (1935).

[83] Ya. I. Frenkel, Usp. Fiz. Nauk, 20, 84 (1938).

[84] K. Karamcheti and L. B. Scott, Jr., J. Chem. Phys., 50, 2364 (1969).

[85] R. M. Logan and R. E. Stickney, J. Chem. Phys., 44, 195 (1966).

[86] R. M. Logan, J. C. Keck, and R. E. Stickney, in Rarefied Gas Dynamics, Proc. 5th Intern. Symp., Oxford, 1966 (C. L. Brundin, ed.), Vol. 1, Academic Press, New York, 1967, p. 49.

[87] R. M. Logan and J. C. Keck, J. Chem. Phys., 49, 860 (1968).

[88] F. O. Goodman, J. Phys. Chem. Solids, 26, 85 (1965).

[89] R. E. Stickney, R. M. Logan, S. Yamamoto, and J. C. Keck, in Proc. Symp. Fundamentals of Gas-Surface Interactions, San Diego, 1966 (H. Saltsburg, J. N. Smith, Jr., and M. Rogers, eds.), Academic Press, 1967, p. 422.

[90] H. Saltsburg and J. N. Smith Jr., J. Chem. Phys., 45, 2175 (1966).

[91] R. E. Forman, J. Chem. Phys., 55, 2839 (1971).

[92] A. U. MacRae, Surface Sci., 2, 522 (1964).

[93] H. B. Lyon and G. A. Somorjai, J. Chem. Phys., 44, 3707 (1966).

[94] D. D. Konowalow and J. O. Hirschfelder, Phys. Fluids, 4, 629 (1961).

[95] G. Ehrlich, in Molecular Processes at the Gas-Solid Interface, Structure and Properties of Thin Films, 1959, Wiley, New York, p. 433.

[96] W. Watt and R. Moreton, RAE (Farnborough) Tech. Note CPM80, 1964.

[97] J. D. McClure, J. Chem. Phys., 51, 1687 (1969).

[98] F. O. Goodman, J. Chem. Phys., 53, 2281 (1970).

[99] J. N. Smith Jr., and H. Saltsburg, in Rarefied Gas Dynamics, Proc. 4th Intern. Symp., Toronto, 1964 (J. H. de Leeuw, ed.), Vol. 2, Academic Press, New York, 1966, p. 491.

[100] R. A. Oman, A. Bogan and C. H. Li, in Rarefied Gas Dynamics, Proc. 4th Intern. Symp., Toronto, 1964 (J. H. de Leeuw, ed.), Vol. 2, Academic Press, New York, 1966, p. 396.

[101] J. D. Kelley and M. Wolfsberg, J. Chem. Phys., 44, 324 (1966).

[102] D. Secrest and B. R. Johnson, J. Chem. Phys., 45, 4556 (1966).

[103] D. Secrest, J. Chem. Phys., 51, 421 (1969).

[104] D. J. Diestler, J. Chem. Phys., 52, 2280 (1970).

[105] B. H. Mahan, J. Chem. Phys., 52, 5221 (1970).

[106] F. O. Goodman, Surface Sci., 7, 391 (1967).

[107] D. F. Jackson and J. B. French, in Rarefied Gas Dynamics, Proc.
       6th Intern. Symp., Cambridge, Mass., 1968 (L. Trilling and
       H. Wachman, eds.), Vol. 2, Academic Press, New York, 1969,
       p. 1119.

[108] M. R. Busby, J. D. Haygood, and C. H. Link, Jr., J. Chem. Phys.,
       54, 4642 (1971).

[109] D. R. O'Keefe and J. B. French, in Rarefied Gas Dynamics, Proc.
       6th Intern. Symp., Cambridge, Mass., 1968 (L. Trilling and
       H. Wachman, eds.), Vol. 2, Academic Press, New York, 1969,
       p. 1279.

[110] R. A. Oman, A. Bogan, C. Weiser, and C. H. Li, AIAA J., 2, 10
       (1964); R. A. Oman, J. Chem. Phys., 48, 3919 (1968); R. A. Oman,
       in Rarefied Gas Dynamics, Proc. 6th Intern. Symp., Cambridge,
       Mass., 1968 (L. Trilling and H. Wachman, eds.), Vol. 2, 1969,
       p. 1331; V. S. Calia and R. A. Oman, J. Chem. Phys., 52, 6184
       (1970).

[111] L. M. Raff, J. Lorenzen, and B. C. McCoy, J. Chem. Phys., 46,
       4265 (1967); J. Lorenzen and L. M. Raff, J. Chem. Phys., 49, 1165
       (1968); J. Lorenzen and L. M. Raff, J. Chem. Phys., 52, 1133
       (1970); J. Lorenzen and L. M. Raff, J. Chem. Phys., 52, 6134
       (1970); J. Lorenzen and L. M. Raff, J. Chem. Phys., 54, 674 (1971).

[112] J. D. McClure and Y. Wu, in Rarefied Gas Dynamics, Proc. 6th
       Intern. Symp., Cambridge, Mass., 1968 (L. Trilling and H. Wachman,
       eds.), Vol. 2, Academic Press, New York, 1969, p. 1191.

[113] J. D. McClure, J. Chem. Phys., 52, 2712 (1970).

[114] F. O. Goodman, Surface Sci., 26, 327 (1971).

[115] J. D. McClure, to be published (cited in Ref. [114]).

[116] J. C. Maxwell, in The Scientific Papers of James Clerk Maxwell,
       Vol. 2, p. 708, Cambridge Univ. Press, 1890.

[117] R. Schamberg, RAND Res. Memo, RM2313 (1959).

[118] G. E. Cook, Planet. Space. Sci., 13, 929 (1965).

[119] F. O. Goodman, in The Structure and Chemistry of Solid Surfaces,
       (G. A. Somorjai, ed.), Wiley, New York, 1969, pp. 40-41.

[120] S. Nocilla, in Rarefied Gas Dynamics, Proc. 3rd Intern. Symp.,
       Paris, 1962 (J. A. Laurmann, ed.), Vol. 1, Academic Press,
       New York, 1963, p. 315.

[121] F. C. Hurlbut and F. S. Sherman, Phys. Fluids, 2, 486 (1968).

[122] S. Nocilla and M. G. Chiado–Piat, in Rarefied Gas Dynamics, Proc. 6th Intern. Symp., Cambridge, Mass., 1968 (L. Trilling and H. Wachman, eds.), Vol. 2, Academic Press, New York, 1969, p. 1069.

[123] I. Kuščer, in Rarefied Gas Dynamics, Proc. 7th Intern. Symp., Pisa, 1970 (to be published).

[124] I. Kuščer, Surface Sci., 25, 225 (1971).

[125] S. F. Shen and I. Kuščer, in Rarefied Gas Dynamics, Proc. 7th Intern. Symp., Pisa, 1970 (to be published).

[126] A. Klavins, in Rarefied Gas Dynamics, Proc. 7th Intern. Symp., Pisa, 1970 (to be published).

[127] C. Cercignani, in Transport Theory, Proc. SIAM–AMS (G. Birkhoff et al., eds.), Vol. 1, Amer. Math. Soc., 1969, p. 249.

[128] E. P. Wenaas, J. Chem. Phys., 54, 376 (1971).

[129] D. R. Miller and R. B. Subbarao, J. Chem. Phys., 55, 1478 (1971).

[130] F. O. Goodman, in Rarefied Gas Dynamics, Proc. 6th Intern. Symp., Cambridge, Mass., 1968 (L. Trilling and H. Wachman, eds.), Vol. 2, Academic Press, New York, 1969, p. 1105.

[131] P. J. Pagni and J. C. Keck, in Rarefied Gas Dynamics, Proc. 7th Intern. Symp., Pisa, 1970 (to be published).

[132] Cited in Ref. [63].

[133] R. J. Madix and R. A. Korus, J. Phys. Chem. Solids, 29, 1531 (1968).

CHAPTER 2

SURFACE SPACE CHARGE LAYERS IN SOLIDS

Daniel R. Frankl

Department of Physics
The Pennsylvania State University
University Park, Pennsylvania

1. INTRODUCTION

The theoretical study of the bulk properties of crystalline solids starts with the assumption of macroscopic electrical neutrality. Thus the microscopic potential is necessarily a periodic function of position, and one obtains the familiar Bloch wave functions and energy bands.* The next step is to consider a variety of perturbations that may be applied. The latter include externally applied electrical and magnetic fields, chemical impurities, atomic displacements, etc., which produce departures from periodicity and alterations of the wave functions and energies. If the perturbation is weak, the alterations are slight and a first-order perturbation theoretic calculation is quite adequate. This is the sort of approach that leads to the usual theories of linear transport such as electrical conductivity and the like. As the perturbation becomes stronger, it obviously becomes necessary either to go to higher orders of perturbation theory or to adopt some different approach. Needless to say, such cases are not yet very thoroughly worked out.

The particular perturbation that is of concern in the present chapter is the capacitively applied electric field. By this, we mean a field arising due to the presence of external charges. It is to be distinguished from a conductively applied field, which is obtained by connecting the sample in direct contact with a source of electromotive force. A static capacitive field does not evoke a steady flow of current. Hence, it is readily possible to make such fields extremely strong.

In the case of nonconducting (i.e., dielectric) crystals, the capacitive field simply polarizes the lattice. There are displacements of the (time-average) positions of charges, but they are small on the scale of the unit cell dimensions. Thus, each unit cell acquires a dipole moment and the material acquires a macroscopic polarization. This tends to weaken the applied field but not truly to screen it out. That is to say, the field penetrates an infinite distance through a dielectric material.

*It is assumed that the reader is familiar with these subjects. They are treated in many texts on solid state physics.

In a conductor, on the other hand, charge carriers are free to move over many unit cells. There can then develop a net macroscopic charge density which can truly screen out the field, so that its effects become vanishingly small beyond a certain finite depth of material. The region in which the field and the net charge density are present is called the "space charge layer." Its thickness depends very much on the available concentration of free carriers, and may range from a few atomic spacings in metals up to microns, centimeters, or even meters in common semiconductors. Needless to say, dielectric weakening of the field is also present within the space charge layer.

An idea of the vast scope of space charge theory can be gleaned by considering the ingredients required for a complete description of the behavior of a medium containing mobile charge carriers. These include:

(1) the Poisson equation, relating the macroscopic potential distribution to the net macroscopic charge density (this equation will be discussed in some detail in Section 2);

(2) the current-flow law for each type of carrier;

(3) the continuity equation for each type of carrier;

(4) boundary conditions for the macroscopic potential;

(5) boundary conditions for each type of carrier; and

(6) expressions for transition rates among the different types of carriers (i.e., "generation" and "recombination" rates).

Clearly, it is hardly to be expected that anything approaching a general solution of this system has been developed. Instead, only special cases, some fairly narrow, some a bit broader, have been treated. A useful, though avowedly nonexhaustive, classification of the latter has been proposed by MacDonald [1]. In this context, the present review is confined almost entirely to the "zero-current, thermal equilibrium" case. Thus items (2) and (3) in the list just given are replaced by a concentration law for each carrier type, and item (6) is omitted. The resulting simplification then makes it possible to be fairly general in treating other aspects of the problem, such as the nature of the material.

Historically, the study of space charge in solids first arose in connection with rectification at metal-to-semiconductor contacts. The early theories [2, 3] recognized the existence of a "barrier layer" at the contact, and attempted to calculate the macroscopic potential distribution within the layer via various simple models. Because the current-flow problem was primary, the treatments of the charge contributions were greatly oversimplified, often to the point of considering only the fixed charges. This early work is well summarized and discussed in the books by Mott and Gurney [4], Henisch [5], and Spenke [6].

The "zero-current" branch of space charge theory had additional roots in the study of chemisorption and catalysis [7-9]. This line of investigation came together with semiconductor surface physics upon the discovery [10] of the "ambient cycling" method of control of the surface potential. The latter method, along with the "field effect" [11], was responsible for opening up the quantitative experimental study of semiconductor surface effects, which in turn stimulated the need for a detailed theory. Although the 1953 paper by Weisz [9] contained the essential ideas, the complete formulation was not made until 1955, independently by Garrett and Brattain [12] and Kingston and Neustadter [13]. These papers dealt with the simplest of the multi-carrier cases, namely a nondegenerate semiconductor with fully ionized impurity levels. Garrett and Brattain included also the simplest of the possible nonequilibrium conditions, namely "quasi-equilibrium" (cf. Ref. [14], p. 308). Various other extensions of the theory will be discussed in later sections. The subject has been covered in considerable detail in books by Many, Goldstein, and Grover [15] and the present author [16].

The purpose of the present chapter is to reexamine the theory from a broader and deeper point of view. An attempt will be made to look into the underlying assumptions and the nature of the approximations entailed, so as to point out avenues of possible future extensions. At the same time, the range of materials from near-insulators to semimetals will be covered in a unified manner. (Metals, unfortunately, are in a class by themselves owing to the overwhelming influence of many-body effects.) Emphasis will be given to areas of rather recent interest, such as quantization effects and space charge capacitance. Of course, some review of older material will be necessary, but will be held to a minimum.

2. THE POISSON EQUATION

As was noted in Section 1, a common ingredient in all studies of space charge is the well known Poisson equation of electrostatics. This is often erroneously regarded as a fundamental and exact equation. That such is not the case can be seen by following through the derivation given in a few advanced texts, such as the one by Jackson [17]. The starting point is the equation that is (presently believed to be) rigorously exact, namely Gauss' law. Since the latter pertains to charges in empty space, it can only be applied to matter on a microscopic (i.e., subatomic) scale, so that it reads*

$$\epsilon_0 \, \text{div} \, \mathscr{E}_{mic} = \rho_{mic} \tag{2.1}$$

If one wishes to avoid the complexity of the rapidly fluctuating microscopic quantities by dealing with the corresponding macroscopic ones, the latter must be defined as suitable averages. Thus,

$$\epsilon_0 \, \text{div} \, \mathscr{E} = <\rho_{mic}> \tag{2.2}$$

where

$$\mathscr{E} \equiv <\mathscr{E}_{mic}> \tag{2.3}$$

The evaluation of $<\rho_{mic}>$ is clearly the crux of the problem. It cannot be done in closed form in general but requires some sort of series expansion. In Jackson's treatment, this is a multipole expansion of $\mathscr{E}_{mic}$. Perhaps somewhat more elegant is a recent treatment by Russakoff [18] involving

*We shall use rationalized mks units throughout. To convert to Gaussian units, set $\epsilon_0 = (4\pi)^{-1}$ and replace D by $D/4\pi$.

the Taylor series expansion of a weighting function used in defining the averaging process. Russakoff's result is, in different notation

$$\epsilon_0 \ \text{div} \ \mathscr{E}(\underline{r}) = \rho(\underline{r}) + \rho_{mol}(\underline{r}) \tag{2.4}$$

where

$$\rho_{mol} = - \text{div} \left[\underline{P} - \text{div} \ \overset{\leftrightarrow}{\underline{Q}} + \cdots \right] \tag{2.5}$$

Here ρ is the average density of "free" or "true" charge. Russakoff defines true charge as the charge of the conduction electrons, but clearly this is too restrictive. In addition to the obvious extension to include all types of carriers, one must, as Jackson's derivation shows, include any net charge of the "molecules."* Hence

$$\rho(\underline{r}) = \rho_{carriers}(\underline{r}) + \sum_i <N_i(\underline{r})> <q_i(\underline{r})> \tag{2.6}$$

where N_i and q_i are the concentration and charge of the ith species of molecules.

Returning to Eq. (2.5), $\underline{P}$ is the "polarization", or average concentration of electric dipole moment, and $\overset{\leftrightarrow}{Q}$ is an analogous quantity that could be called the "quadrupolarization," i.e., the average concentration of electric quadrupole moment. The term in $\overset{\leftrightarrow}{Q}$ is usually negligible and rarely even written, but since extremely strong field gradients can exist in space charge layers, it could conceivably be significant in some cases.

From Eqs. (2.4) and (2.5) we obtain
$$\text{div} \ \underline{D} = \rho \tag{2.7}$$
where

$$\underline{D} = \epsilon_0 \mathscr{E} + \underline{P} - \text{div} \ \overset{\leftrightarrow}{\underline{Q}} + \cdots \tag{2.8}$$

In the usual linear response approximation, P is taken to be proportional to $\mathscr{E}$ (and div $\overset{\leftrightarrow}{Q}$ and higher terms are neglected) so that

$$\underline{D} = \kappa\epsilon_0\mathscr{E} = - \kappa\epsilon_0 \ \text{grad} \ V \tag{2.9}$$

*In the case of crystalline media, it would probably be better to use the term "unit cells" instead of "molecules."

and Eq. (2. 7) then is

$$\nabla^2 V = - \rho/\kappa\varepsilon_0 \tag{2.10}$$

which is, of course, the Poisson equation. The purposes of the foregoing somewhat extended discussion are: (1) to emphasize the restrictions and approximations involved, and (2) to make clear just which charges are to be included in ρ, namely those given in Eq. (2.6).

3. THE CHARGE DENSITY

Clearly the first job is to evaluate the charge density, Eq. (2.6), as function of the variables that determine it. Consider first the free carrier term. It is clear that this is related to the wave functions of the appropriate conduction and valence band states. Now, for the macroscopic charge density the detailed variations within each unit cell are unimportant; only the variation in amplitude from cell to cell need be known. However, this must be known under the following conditions: (1) the boundary conditions pertinent at the actual surface, and (2) the presence of the macroscopic field. For example, the familiar Bloch functions having constant amplitude from cell to cell are the correct wave functions only in zero field and under periodic boundary conditions.

Let us first consider the effects of the boundary conditions. Most commonly, the conductor is in contact with vacuum or an insulator, so that the lowest available outside energy level lies considerably (say several eV) above the highest occupied inside level. Then it would not be too bad an approximation to regard the surface as an infinite potential step and require the wave function to go to zero. The wave function will then take the form

$$\psi_n(\underline{k}, \underline{r}) = U_n(\underline{k}, \underline{r}) \exp^{i(k_x x + k_y y)} \sin k_z z \tag{3.1}$$

where the $U_n(\underline{k}, \underline{r})$ are the same periodic factors as appear in the Bloch functions, except that k_z is now given by multiples of π/L_z instead of $2\pi/L_z$.

The effect, therefore, is to keep the carriers some distance away from the surface. The distance will be of the order of the wavelength $(2\pi/k_z)$ of the typical occupied state. For a highly degenerate electron gas as in a metal, it will be only a few lattice spacings [19-21], and for a nondegenerate distribution near room temperature it will be some tens of angstroms. Since the space charge layer thickness is ordinarily (i. e., when the field is not too strong) much larger than this, the boundary repulsion effect can usually be neglected. It is entirely possible, however, as we shall see later, for the field to be strong enough to make the layer thickness comparable with the carrier wavelength. It is then that "quantization" effects become significant and that the boundary conditions must be taken into account. These matters will be discussed in Section 5.

We next consider the effect of the field on the wave functions. A great deal of insight into this problem is given by the recent work of Rabinovitch and Zak [22] who extended the Mathieu problem [23, 24] to include fields and boundaries. The result, at least for states near the band edges, is that the envelope function is changed from $\sin k_z z$ to a function appropriate for the tilted potential well (Airy functions in their case of a uniform field), whereas the density of states is not seriously affected. These conclusions do, of course, lose validity when the macroscopic field becomes comparable with the periodic crystal field, but it is doubtful if this condition is ever encountered in practice. For the relatively weak fields of ordinary space charge layers, the form of the envelope function is of no consequence, since the length scale of its variations is, for most of the occupied states, small compared with the scale of variation of the macroscopic field. Thus, it is a valid approximation to deal with wave functions that have uniform amplitudes but whose energies vary additively with the local macroscopic potential energy. For stronger fields, of course, the envelope function must be considered, and one is then in the quantization regime.

Having now described the wave functions (in a preliminary sense at least), it remains to consider the occupancy factors. In the most general case, the latter are governed by kinetic considerations: the occupancy of any given state depends upon those of all other states and upon the rates of transitions to and from them. But since we are considering (mainly) systems in thermal equilibrium, the situation is vastly simplified. All net transition rates are zero, and the occupancies are governed by statistical considerations alone. Undoubtedly this subject is very familiar to most readers, but we shall review it briefly so as to have some formulas on record for later use.

Since electrons have spin 1/2, each spatial wave function generates two quantum states, one with each of the two possible spin orientations. The Pauli exclusion principle permits either zero or one electron to occupy any quantum state at any instant. Now, the occupation of any state will, via the Coulomb repulsion, affect the energies of all other states to some extent. This is just another way of saying that we are really dealing with a many-body problem. However, for most problems involving relatively low carrier concentration (e.g., as in semiconductors, usually) it is adequate to consider only the two extreme cases: (1) band states (nonlocalized), and (2) highly localized states.

For band states, the charge distribution is spread so thin that the Coulomb repulsion may be totally neglected, even for the companion state having the same spatial wave function but opposite spin. Thus all states are occupied independently and the average occupancies are given by the Fermi-Dirac distribution function

$$f_j = [1 + \exp \beta (E_j - E_F)]^{-1} \tag{3.2}$$

If the distribution of energies of the states in a band is known, it is just a matter of integration to determine the total occupancy of the band. These matters are discussed in great detail by Blakemore [25] and many other authors. The principal results that we shall need later pertain to bands of standard form, i.e., quadratic dependence of energy on the components of the wave vector. For the electron or hole concentrations in such bands, we have

$$n = N_C \mathscr{F}_{1/2} (\beta [E_F - E_C]) \tag{3.3}$$

$$p = N_V \mathscr{F}_{1/2} (\beta [E_V - E_F]) \tag{3.4}$$

Here N_C and N_V are the effective concentrations of conduction and valence band states, given by

$$N_C = 2(2\pi m_C kT/h^2)^{3/2}$$

$$N_V = 2(2\pi m_V kT/h^2)^{3/2} \tag{3.5}$$

and $\mathscr{F}_{1/2}$ is one of the Fermi-Dirac integrals

$$\mathscr{F}_j(\xi) = \frac{1}{\Gamma(j+1)} \int_0^\infty \frac{x^j dx}{1 + \exp(x - \xi)} \tag{3.6}$$

Series expansions of these functions are given by Blakemore [25], Wilson [26], and with extensive references to the related mathematical literature, by Hill [27]. For our purposes, it suffices to quote Blakemore's approximation formulas:

$$\mathscr{F}_{1/2}(\xi) \simeq e^\xi (1 + 0.27 e^\xi)^{-1} \qquad \text{for } \xi \leq 1 \tag{3.7}$$

$$\mathscr{F}_{1/2}(\xi) \simeq (4/3\pi^{1/2}) (\xi^2 + 1.7)^{3/4} \qquad \text{for } \xi \geq 1 \tag{3.8}$$

Indeed, in most cases we shall assume either extreme degeneracy ($\xi \gg 1$) or extreme nondegeneracy ($\xi \ll -1$) and drop the second terms in each of these.

We now turn to the evaluation of the second term in the charge density, Eq. (2.6). Since the unit cells of the pure crystal with no macroscopic field present must be electrically neutral, the only contribution will be from those cells containing ionized defect centers. Any given type of defect has a specific range of possible charge states in which it may exist. The charges may be altered by capture (or release) of electrons from (or to) the band states. When a given imperfection is in a particular one of its charge states, it presents a specific set of energy levels into which the next electron can be captured. When this occurs in any one of the set of levels, the charge value of the imperfection becomes one unit more negative and an entirely different set of levels is then presented. Owing to the Coulomb repulsion, the sets of levels tend toward higher and higher energies, until finally no more electrons can be held. Thus, to each charge value (say s electrons) there corresponds some number of quantum states $\psi_{s,j}$ with energies $E_{s,j}$. The problem of the occupation statistics of such sets of states was worked out by Shockley and Last [28]. The result may be stated concisely as

$$\frac{N_{s+1}}{N_s} = \frac{Z_{s+1}}{Z_s} \exp \beta E_F \tag{3.9}$$

where Z_s is the partition function of the set of states $\psi_{s,j}$, viz.,

$$Z_s = \sum_j \exp (-\beta E_{s,j}) \tag{3.10}$$

The foregoing result is readily understandable intuitively as an extension of the ordinary Fermi-Dirac distribution for single states, which may be written

$$\frac{N_{occupied}}{N_{empty}} = \frac{\exp (-\beta E)}{(1)} \exp \beta E_F \tag{3.11}$$

The analogy may be pushed even further by introducing a fictitious energy defined as

$$E'_{s+1,s} \equiv kT \, \ln (Z_s/Z_{s+1}) \tag{3.12}$$

The fraction of the defects in charge states s is, of course,

$$f_s = \frac{N_s}{\sum_s N_s} \tag{3.13}$$

In most cases of interest, only the lowest of the energies $E_{s,j}$ make any significant contribution to Z_s. There may, however, be several states with the same energy (i.e., degeneracy), and in this case

$$\frac{Z_{s+1}}{Z_s} \simeq \frac{g_{s+1}}{g_s} \exp \left[-\beta (E_{s+1} - E_s) \right] \tag{3.14}$$

where the g's are the ground-state degeneracies and the E's the corresponding energies of the defect in the respective charge conditions. (It should be emphasized that the energies refer to the defect alone, not to the crystal as a whole. Thus, removal of an electron from a donor lowers the energy of the donor; if the electron ends up in a higher level, such as a conduction band state, then the energy of the crystal has been raised. As Shockley and Last point out, it is helpful to think of the zero of energy as that of the defect with all its removable electrons removed.)

A few examples may help to clarify these matters.

(1) Simple Donors. These are defects (often substitutional impurities) that may exist with charges $s = 0$ (neutral) or $s = -1$ (positively ionized). In the ionized condition, all valence bonds are complete and electrons are paired, so there is no degeneracy; i.e., $g_{-1} = 1$. The neutral state may contain one electron of either spin, hence $g_0 = 2$. We denote the ground-state energy as E_D, so $E_C - E_D$ is the ionization energy of the donor. Hence, Eqs. (3.14) and (3.9) are

$$\frac{Z_0}{Z_{-1}} = \frac{2}{1} \exp(-\beta E_D) \tag{3.15}$$

and

$$\frac{N_D^0}{N_D^+} = 2 \exp \beta(E_F - E_D) \tag{3.16}$$

From the latter, since only the two charge states may exist, the fraction of ionized donors is

$$f_D^+ \equiv \frac{N_D^+}{N_D} \equiv \frac{N_D^+}{N_D^+ + N_D^0} = [1 + 2 \exp \beta(E_F - E_D)]^{-1} \tag{3.17}$$

(2) Simple Acceptors. These may exist in states of charge $s = 0$ (neutral) or $s = 1$ (negatively ionized). The latter has all electrons paired, hence $g_1 = 1$. The energy of this state is denoted as E_A, so that $E_A - E_V$ is the ionization energy (energy required to remove a hole to the valence band) of the acceptor. The neutral state is missing one electron, which may be of either spin. Hence, if the spatial wave function is nondegenerate we will have $g_0 = 2$. These considerations lead to an equation for the fraction of ionized acceptors quite analogous to Eq. (3.17), namely,

$$f_A^- \equiv \frac{N_A^-}{N_A} \equiv \frac{N_A^-}{N_A^- + N_A^0} = [1 + 2 \exp \beta(E_A - E_F)]^{-1} \qquad (3.18)$$

(3) <u>Defect Levels Near Degenerate Bands</u>. Since the defect levels are mainly drawn out of the adjacent bands [29], degeneracy in the latter may well be reflected in additional degeneracy in the former. This seems to be the case for shallow acceptor levels in germanium and silicon. It was predicted by Kohn [30] and verified by Blakemore [31] at least for germanium, that $g_0 = 4$ rather than 2. Thus we have in this case

$$\frac{Z_1}{Z_0} = \frac{1}{4} \exp(-\beta E_A) \qquad (3.19)$$

$$\frac{N_A^-}{N_A^0} = \frac{1}{4} \exp \beta(E_F - E_A) \qquad (3.20)$$

and

$$f_A^- = [1 + 4 \exp \beta(E_A - E_F)]^{-1} \qquad (3.21)$$

The reader is reminded that in these examples we have neglected the excited states of the defects. The conditions under which this is valid are discussed at some length by Blakemore [25, pp. 140ff]. Clearly the error will be greatest in the temperature range where the energy level spacings are comparable with kT, since at higher temperatures all the defects are ionized anyway. Some examples cited suggest that errors of as much as 50 percent in carrier concentration may occur. Unfortunately, there is rarely enough information available to permit an exact calculation, especially inasmuch as the defects interact with each other so that the energies are concentration-dependent. This does tend, though, to alleviate the difficulty at the higher concentrations, since the excited states get pushed into the bands first. Of course, when all the impurity states get pushed into the bands, the considerations regarding localized states no longer hold. Instead, the levels are independently occupied just like the band states.

The question of the density-of-states function for the band states under the latter conditions requires a bit of discussion. At low concentrations of defects (say $N_D \ll N_C$ or $N_A \ll N_V$) the defect states are discrete and are to be regarded as having been perturbed out of the bands. Thus, strictly speaking $N_C - N_D$ and $N_V - N_A$ should be taken as the effective concentrations of band states, but the corrections are negligible at low defect concentrations. As the concentrations are increased, it is roughly when N_D becomes comparable with N_C (or N_A with N_V) that the states re-merge with the bands. In other words, N_C and N_V are then the correct combined effective state concentrations. To a good approximation, the band shape remains unchanged. But when the defect concentrations are increased still further, the band-edge energies and densities of states are significantly affected. This is evident experimentally in a variety of ways as has been discussed by Pankove [32] for germanium. No comprehensive theory exists as yet, though Conwell and Levinger [33] have carried out a perturbation calculation for germanium that gives a reasonably good value for the gap shrinkage. A recent paper by Kaplan [34] predicts a bulge in the density of states that broadens and moves into the band with increasing concentration.

Assembling the foregoing results, we find for the second term of Eq. (2.6)

$$\sum_i\ <N_i>\ <q_i>\ =\ e\left[N_D f_D^{\ +} - N_A f_A^{\ -} - \sum_{t,s} N_t s f_{t,s}\right] \qquad (3.22)$$

where, in the last term, the subscript t designates the various types of deep-level defects (traps) present. All of the quantities in the brackets may, of course, be functions of position in the crystal, although we shall consider in this chapter only cases where the concentrations of defects are uniform. The spatial averages on the left side of Eq. (3.22) are understood to be taken over regions large compared with the spacing of the defects but small compared with the distance in which the occupancy factors vary appreciably. Needless to say, these conditions could be incompatible.

One thing remains to complete the description of the various contributions to the macroscopic charge density (at equilibrium). That is to specify the value of the Fermi energy E_F, which appears as a parameter in all the occupancy factors. This can always be done by noting that any region far from all surfaces (and from all changes in defect concentration, if there be any) must be electrically neutral. A few examples will serve to illustrate how this works.

Intrinsic Materials. If there are no charged defects present in any appreciable concentration, then only the band states need be considered. Assuming the standard form, Eqs. (3.3) and (3.4) then give the neutrality condition as

$$-N_C \mathscr{F}_{1/2}(\beta[E_i - E_C]) + N_V \mathscr{F}_{1/2}(\beta[E_V - E_i]) = 0 \qquad (3.23)$$

where

E_i = intrinsic Fermi energy

= value of E_F in the bulk of an intrinsic sample of the material

For ordinary semiconductors, where $E_C - E_V \gg kT$, the approximation Eq. (3.7) will hold and, neglecting the second term of that approximation, the solution of Eq. (3.23) is

$$E_i = \frac{1}{2}(E_C + E_V + kT\ln N_V/N_C) \qquad (3.24)$$

In this case, the electron and hole concentrations are each equal to

$$n_i = N_C \exp(-w_C) = N_V \exp w_V = (N_C N_V)^{1/2} \exp(-\beta E_g/2) \qquad (3.25)$$

where $w_C = \beta(E_C - E_i)$, $w_V = \beta(E_V - E_i)$

The opposite extreme is $E_C - E_V \ll kT$. This can occur, conceivably, for a narrow-gap semiconductor at high temperature, but is more characteristic of a semimetal, for which $E_C - E_V$ is negative (i.e., the bands overlap). The approximation Eq. (3.8), again with the neglect of the second term, then gives

$$\left(\frac{E_i - E_C}{E_V - E_i}\right)^{3/2} = \frac{N_V}{N_C} = \left(\frac{m_V}{m_C}\right)^{3/2}$$

or

$$E_i = \frac{m_V E_V + m_C E_C}{m_V + m_C} \qquad (3.26)$$

The carrier concentrations now are

$$n_i \equiv \frac{4}{3\pi^{1/2}} N_C (-w_C)^{3/2} \tag{3.26}$$

$$= \frac{4}{3\pi^{1/2}} N_V w_V^{3/2} \tag{3.27}$$

Extrinsic Materials. Usually the only defects present in statistically significant concentrations are the shallow donors and acceptors. In this case, the neutrality condition reads

$$N_C \mathscr{F}_{1/2}(\beta[E_F - E_C]) - N_V \mathscr{F}_{1/2}(\beta[E_V - E_F]) = N_D f_D^+ - N_A f_A^- \tag{3.28}$$

The general result that $E_F > E_i$ if $N_D > N_A$, and vice versa, is undoubtedly familiar to the reader. Quantitative solutions of Eq. (3.28) under a wide variety of conditions of impurity concentrations and temperature are discussed by Blakemore [25] and have been summarized in the texts [15, 16]. The only formula we shall repeat here is for the case that is by far the most common in practice, namely, fully ionized donors and acceptors, and of course nondegenerate statistics in the bands. In this case, the Fermi energy is given by

$$u \equiv \beta(E_F - E_i) = \sinh^{-1} \frac{N_D - N_A}{2 n_i} \tag{3.29}$$

and the carrier concentrations by

$$n = n_i \exp u$$

$$p = n_i \exp(-u) \tag{3.30}$$

The case of semimetals (for which $E_V \geq E_C$) containing donors or acceptors is less familiar, but has been discussed by Boyle and Smith [35] from the experimental point of view. In the elemental semimetals, donors (i.e., substitutional impurities with one more valence electron than the host) raise the Fermi energy and acceptors lower it, just as in semiconductors. It is perhaps not obvious why this should be so since the donor levels (say) should be well below the Fermi level, so that all the donors would be expected to be neutral. However, since the defects can be

statistically significant only if they are present in concentrations at least
fairly comparable with the intrinsic concentration, it is likely that the levels
will be merged with the bands. Thus, as a first approximation it would be
reasonable to take the combined density of states to be the same as the unperturbed band density. Then the neutrality condition reads, under the approximation given by the first term of Eq. (3.8),

$$\frac{4}{3\pi^{1/2}} \left[\frac{N_C}{(kT)^{3/2}} (E_F - E_C)^{3/2} - \frac{N_V}{(kT)^{3/2}} (E_V - E_F)^{3/2} \right] = (N_D - N_A) \quad (3.31)$$

The carrier concentrations then are, assuming Eq. (3.31) to have been
solved for E_F,

$$n = n_i \left(\frac{E_F - E_C}{E_i - E_C} \right)^{3/2} = n_i \left(\frac{u - w_C}{-w_C} \right)^{3/2} = n_i \left(\frac{|w_C| + u}{|w_C|} \right)^{3/2} \quad (3.32a)$$

$$p = n_i \left(\frac{E_V - E_F}{E_V - E_i} \right)^{3/2} = n_i \left(\frac{w_V - u}{w_V} \right)^{3/2} \quad (3.32b)$$

4. THE SPACE CHARGE LAYER: NONQUANTAL REGIME

The formulas of the latter part of Section 3 relating the Fermi energy to
the band-edge energies pertain, as was stated there, only to regions sufficiently remote from surfaces and other discontinuities. We shall refer to
such regions as bulk regions of the material. Now the Fermi energy, outside of the bulk regions, is the same (in equilibrium) as inside since it is
the electrochemical potential of the electrons. Therefore, all the energy
levels are shifted in proportion to the local value of the macroscopic electrostatic potential. In other words, taking the bulk value as the reference
level for potential,

$$eV(\underline{r}) = E_{CB} - E_C(\underline{r}) \quad (4.1a)$$

$$= E_{VB} - E_V(\underline{r}) \quad (4.1b)$$

$$= E_{iB} - E_i(\underline{r}) \quad (4.1c)$$

where the subscript B denotes the bulk value of each energy. Thus, the contributions to the charge density will, in general, vary with position in the crystal. Two main cases may be distinguished: in the nonquantal regime, the variation of V is sufficiently slow that each region behaves as a bulk region with the same energy levels ($E_C - E_F$, etc.). In this case, the charge density is a function of the local potential, and the Poisson equation, Eq. (2.10), governs this relationship. In the quantal regime, where the potential varies more rapidly, the charge density is a functional of the entire potential distribution, governed by the Schrödinger equation. We shall consider this case in the next section.

The Poisson equation is of the general form

$$\nabla^2 V = - (\kappa \varepsilon_0)^{-1} \rho (V, \; E_F - E_{iB}) \tag{4.2}$$

where the single parameter $E_F - E_{iB}$ designates the dependence on the defect concentrations. Of course, the equation cannot in general be solved analytically. However, if the variation of V takes place in one dimension only (i.e., in an infinite plane-parallel slab), one quadrature can always be carried through, because then

$$\nabla^2 V = \frac{d^2 V}{dz^2} = \frac{dV}{dz} \frac{d}{dV} \left(\frac{dV}{dz} \right) = \frac{1}{2} \frac{d}{dV} (\mathscr{E}^2) \tag{4.3}$$

so that

$$\mathscr{E}^2 (z) - \mathscr{E}^2 (z_0) = - \frac{2}{\kappa \varepsilon_0} \int_{V(z_0)}^{V(z)} \rho \; dV \tag{4.4}$$

The question of whether or not a one-dimensional variation is physically realistic will be taken up in Section 7.

4.1. BOUNDARY CONDITIONS

The last equation may be subject to a variety of types of boundary conditions depending on the physical situation under analysis. The simplest occurs at great depths in the material and thus is pertinent if the sample is

thick enough. It states simply that all effects due to the surface must die
away, i.e.,

$$\left.\begin{array}{l} \mathscr{E}(z_0) \to 0 \\[1em] V(z_0) \to 0 \end{array}\right\} \quad \text{as } z_0 \to \infty \qquad (4.5)$$

Boundary conditions at the surface are much more complicated. To discuss them, we must first define a quantity called the vacuum energy, E_{vac}:

$$E_{vac} \equiv \begin{array}{l} \text{energy of an extra electron outside the material} \\ \text{beyond the range of the image force.} \end{array} \qquad (4.6)$$

E_{vac} may be related to the interior energy levels via either of two parameters; the work function ϕ or the electron affinity χ. These are defined as
follows:

$$\phi = E_{vac} - E_F \qquad (4.7)$$

$$\chi = E_{vac} - E_{CS} \qquad (4.8)$$

where $E_{CS} \equiv E_C(z_S)$ = conduction band edge energy at the surface. Both of
these parameters vary with the surface conditions but not in the same way:
χ depends upon dipole layers outside the surface, but not on band bending
within, whereas ϕ depends on both. The subject of work functions and electron affinities is a vast one in itself. A recent review is given by Rivière
[36] with primary emphasis on metals. Data on some semiconductors are
given by Geppert et al. [37]. We shall find it convenient for later discussion to designate an energy $E^*(z) = E_C(z) + \chi$. The various quantities are
illustrated in Fig. 4.1.

Now, a point that is often not fully appreciated is that E_{vac} may be different for different samples. For example, clean tungsten is known [36]
to have appreciably different ϕ values on different crystallographic faces,
so that work would be done in taking an electron "around a corner" outside
a single tungsten crystal. The difference between E_{vac} values for two
samples in equilibrium (i.e., at the same temperature and electrically
connected) is what gives rise to the contact potential difference. This

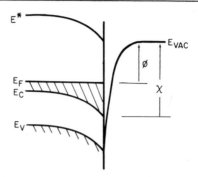

FIG. 4.1. Surface region of a semiconductor (degenerate n-type) showing the band bending, the image potential, and the definitions of work function ϕ and electron affinity χ.

produces an electrostatic field in the space between the materials, and as
they are brought into contact the field influences the band bending in each.
When the surfaces finally touch there can, of course, be no remaining potential difference since the field would be infinite. Hence, one boundary
condition at any interface is

$$E^*_{S,1} = E^*_{S,2} \qquad\qquad (4.9a)$$

or

$$E_{CS,1} + \chi_1 = E_{CS,2} + \chi_2 \qquad\qquad (4.9b)$$

or, with reference to Eq. (4.1a),

$$E_{CB,1} - eV_{S,1} + \chi_1 = E_{CB,2} - eV_{S,2} + \chi_2 \qquad\qquad (4.9c)$$

The situation for two semiconductors in contact is illustrated in Fig. 4.2,
which, for clarity, is drawn on the assumption of no band bending in either
material when they are separated.

A second boundary condition is given in terms of the macroscopic electric field strength by the application of Gauss' Law to a "pillbox" enclosing
the interface. Taking the positive sense of the z axis as the direction from
material 1 to material 2, we obtain

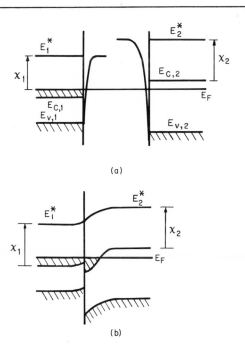

(a)

(b)

FIG. 4.2. Surface regions of two semiconductors having initially flat bands: (a) separated but electrically connected; (b) brought into contact. In the example shown, material 2 is nondegenerate n-type, and acquires a degenerate accumulation layer on contact.

$$\kappa_2 \mathscr{E}_{Z,2}(z_S^+) - \kappa_1 \mathscr{E}_{Z,1}(z_S^-) = Q_{int}/\varepsilon_0 \qquad (4.10)$$

where Q_{int} is the total interfacial charge density (per unit area). This includes electronic charges trapped in surface states in the two materials as well as fixed (e.g., ionic) charges at the interface, so that

$$Q_{int} = Q_{SS}(V_S) + Q_0 \qquad (4.10a)$$

The application of both boundary conditions to a semiconductor-insulator interface is illustrated in Fig. 4.3. It should be borne in mind that because Q_{int} generally depends on V_S, the boundary conditions are not usually explicitly applicable.

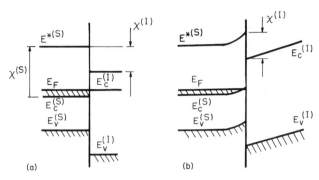

FIG. 4.3. Contact between a semiconductor (degenerate n-type) and an insulator with no interface charge: (a) no field in insulator; (b) with field in insulator.

4.2. SOLUTIONS FOR THICK SAMPLES

If the sample extends far enough for bulk conditions to be (very nearly) attained somewhere within it, then the two surfaces are independent, and each may be analyzed by taking $z_0 \to \infty$ in Eq. (4.4). Thus

$$\mathscr{E}^2(V) = - \frac{2}{\kappa \varepsilon_0} \int_0^V \rho(V')dV' \tag{4.11}$$

$\rho(V')$ is obtained by putting the local energies, given by equations like Eqs. (4.1), into the various expressions from Section 3. To summarize, the latter are

$$\rho_n(z) = - e\, n(z)$$

$$= - e\, N_c \mathscr{F}_{1/2}(\beta[E_F - E_{CB} + eV(z)]) \tag{4.12}$$

$$\rho_p(z) = e\, N_V \mathscr{F}_{1/2}(\beta[E_{VB} - E_F - eV(z)]) \tag{4.13}$$

$$\rho_D(z) = e\, N_D \left| 1 + \exp \beta[E_F - E_{DB}' + eV(z)] \right|^{-1} \tag{4.14}$$

$$\rho_A(z) = - e\, N_A \left| 1 + \exp \beta[E_{AB}' - E_F - eV(z)] \right|^{-1} \tag{4.15}$$

Here, the "effective energies" of the defects are given by Eq. (3.12). For ordinary donors and acceptors in germanium and silicon, in the temperature range where the excited states of the defects may be neglected, these are, from Eqs. (3.17) and (3.21),

$$E_D' = E_D - kT \ln 2 \tag{4.16}$$

$$E_A' = E_A + kT \ln 4 \tag{4.17}$$

The bulk band-edge energies are given in terms of the defect concentrations by Eq. (3.28) in the general case or Eq. (3.29) in the familiar simple case.

4.2.1. Simple Case (Nondegenerate)

The case of a semiconductor with the Fermi level everywhere well within the gap has been thoroughly discussed. The theory was originally developed independently by Garrett and Brattain [12] and by Kingston and Neustadter [13], and various useful approximations were derived by the present author [38], Goldberg [39], Flietner [40], Lee and Mason [41], and Berz [42]. The entire subject is covered in detail in the texts [15, 16] and will only be briefly summarized here.

The $\mathscr{F}_{1/2}$ functions in Eqs. (4.12) and (4.13) are approximated by the first term of Eq. (3.7) so that we obtain, with the aid of Eq. (3.25),

$$n(z) = n_i \exp \beta [E_F - E_{iB} + eV(z)] \tag{4.18a}$$

$$= n_i \exp u(z) \tag{4.18b}$$

$$p(z) = n_i \exp [-u(z)] \tag{4.19}$$

Also, the curly-bracketed factors in Eqs. (4.14) and (4.15) are each unity, so that, in view of Eq. (3.29),

$$N_D^+ - N_A^- = N_D - N_A$$

$$= n_i (\exp u_B + \exp [-u_B]) \tag{4.20}$$

Thus, Eq. (4.11) becomes, since $dV = (kT/e)du$,

$$\mathscr{E}^2(z) = \frac{2n_i kT}{\kappa \varepsilon_0} F^2(u, u_B) \tag{4.21}$$

where

$$|F(u, u_B)| \quad \left[\int_{u_B}^{u} (e^{u'} - e^{-u'} - e^{u_B} + e^{-u_B}) du' \right]^{1/2} \tag{4.22a}$$

$$= \left\{ [(e^u - e^{u_B}) + (e^{-u} - e^{-u_B}) + 2(u_B - u) \sinh u_B] \right\}^{1/2} \tag{4.22b}$$

$$= \left\{ [\Delta n + \Delta p - \Delta u (N_D - N_A)]/n_i \right\}^{1/2} \tag{4.22c}$$

The latter form exhibits the separate contributions of the electrons, the holes, and the ionized defects.

Since $\mathscr{E}_Z = - dV/dZ$, we may eliminate the sign ambiguity by writing

$$\mathscr{E}_Z = - \left(\frac{2n_i kT}{\kappa \varepsilon_0} \right)^{1/2} F(u, u_B) \tag{4.23}$$

where

$$F = - \, \text{sgn}(u - u_B) |F| \tag{4.24}$$

and sgn denotes the sign of its argument. Various types of graphical representations of the function $F(u, u_B)$ have been given. One example is shown in Fig. 4.4. This may be extended to negative u_B values by means of the identity

$$F(-u, -u_B) = - F(u, u_B) \tag{4.25}$$

It is seen that strong (and rapidly varying) fields are encountered when $|u - u_B| \gg 0$. When $(\text{sgn } u_B)(u - u_B) \gg 0$, the dominant space charge contribution comes from the majority carriers, and the space charge layer is termed an accumulation layer. These appear in the lower right-hand corner of Fig. 4.4. When $(\text{sgn } u_B)(u - u_B) \ll 0$, the bulk minority carriers are in the majority in the surface layer and make the dominant space charge contribution. This is termed an inversion layer, and appears in the upper left-hand corner of Fig. 4.4. Between the bulk and the inversion layer, there is a depletion layer, where there are few carriers of either sign. The dominant space charge there is due to the ionized defects. This uniform space charge density leads to the familiar parabolic potential distribution, which appears in Fig. 4.4 as the plateau regions in the upper half.

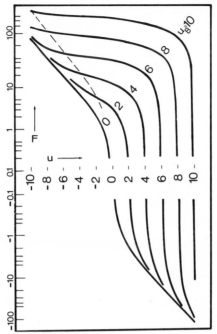

FIG. 4.4. The space charge function $F(u, u_B)$ defined in Eqs. (4.24) and (4.22b). The region to the left of the dashed curve corresponds to inversion, the region to the right to depletion, and the bottom half to accumulation.

In passing, it is worth noting that the existence of a depletion layer between the bulk and an inversion layer is responsible for many of the interesting and useful properties of the inversion layer.

The shape of the potential distribution is obtained by integrating the relation $dz = dV/\mathscr{E}_z(V)$. Formally we obtain

$$z_2 - z_1 = -L_D \int_{u_1}^{u_2} \frac{\kappa du}{F(u, u_B)} \tag{4.26}$$

where

$$L_D = \left(\frac{\epsilon_0 kT}{2e^2 n_i} \right)^{1/2} \tag{4.27}$$

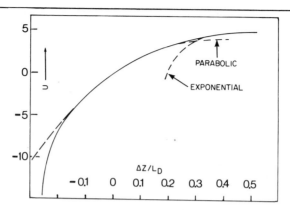

FIG. 4.5. Illustration of the space-charge potential distribution for
$u_B = 5$. (After Ref. [16], Fig. 214.1.)

so that

$$F \equiv L_D \, du/dz \qquad\qquad (4.27a)$$

L_D is a characteristic length called the Debye length or intrinsic screen-
ing length. The foregoing integral can be evaluated only in a few special
cases. One of these is the depletion layer which has, as mentioned above,
a quadratic dependence of V on z. Another important case is the bulk
limit, $|u - u_B| \ll 1$. Then a Taylor expansion of Eq. (4.22b) yields
$F = - (\cosh u_B)^{1/2}(u - u_B)$ which results in an exponential decay of $|u - u_B|$
with characteristic length L_B given by

$$L_B = L_D (\cosh u_B)^{-1/2} \qquad\qquad (4.28)$$

This relation provides the criterion for deciding whether a sample is thick
or not. If the total thickness beyond any depletion layer that may be present
exceeds a few times L_B, the sample is thick enough for the foregoing results to apply.

Numerical evaluations of Eq. (4.26) have been presented by Dousmanis
and Duncan [43], Young [44], and Goldberg [39] among others. The example given in Fig. 4.5 is instructive in that it shows how the exponential
and parabolic regions fit onto the complete curve. It also illustrates the
very rapid variation of potential with distance in the inversion region; the
same would also be true in a strong accumulation layer. Lee and Mason

[41] give approximate analytic forms for the barrier shape in various ranges, but unfortunately the latter are rather limited.

The total charge in the space charge layer is readily obtained from Gauss' Law as

$$Q_{SC} = - \kappa \varepsilon_0 \mathscr{E}_z(z_S^+) \tag{4.29a}$$

$$= (2n_i \kappa \varepsilon_0 kT)^{1/2} F(u_S, u_B) \tag{4.29b}$$

$$= 2en_i L_D F(u_S, u_B) \tag{4.29c}$$

where $u_S \equiv u(z_S)$ = value of u at the surface. The contributions of the two types of free carriers to the total charge are readily obtained by integrating the local departures from the bulk concentrations over the thickness of the layer, viz.,

$$\Delta P \equiv \int_0^\infty (p - p_B) dz$$

$$= n_i L_D G(u_S, u_B) \tag{4.30}$$

and

$$\Delta N = n_i L_D G(-u_S, -u_B) \tag{4.31}$$

where

$$G(u_S, u_B) = \int_{u_S}^{u_B} \frac{e^{-u} - e^{-u_B}}{F(u, u_B)} du \tag{4.32}$$

An idea of the behavior of the function G can be gained from the contour map shown in Fig. 4.6. The labelling of the various sectors identifies the types of surface layers. Many et al., in Fig. 4.10 of their text [15], present the picture somewhat differently in terms of four separate functions closely related to G. For fairly accurate computations, Ref. [16] contains a four-place table for limited ranges of the variables, as well as approximation formulas and graphs [38] to go beyond these ranges. Accurate values of F are also obtained from the table by use of the identity

$$G(u_S, u_B) - G(-u_S, -u_B) = 2F(u_S, u_B) \tag{4.33}$$

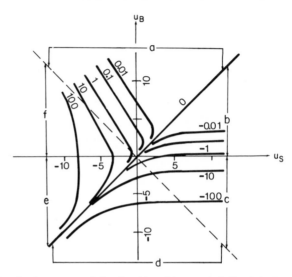

FIG. 4.6. Contour map of the function $G(u_S, u_B)$ defined in Eq. (4.32).
Sectors a and d are the depletion region, b and e accumulation, and c and f
inversion. The upper sectors pertain to bulk majority carriers, the lower
to minority.

Berz [42] gives a set of useful approximations for the derivatives of G and
F with respect not only to u_S but also to the degree of departure from equilibrium.

    To conclude the discussion of the simple case it is interesting to esti-
mate the maximum field strength that may be encountered. Since the requirement is that E_F lie well within the defect levels, the limits on u_S
and u_B are rather vague. For simplicity, we shall extend these somewhat and assume one band-edge in the bulk and the other one at the sur-
face to lie right at the Fermi level. Thus $|u_S - u_B| = E_G/kT$ and, roughly,
$|u_B| = E_G/2kT$. A suitable approximation to Eq. (4.22) for these values
is

$$|F| \simeq (|u_B| \exp |u_B|)^{1/2} \qquad\qquad (4.34)$$

Now, the field coefficient in Eq. (4.23) takes the form

$$\left(\frac{2n_i kT}{\kappa\epsilon_0}\right)^{1/2} \propto \kappa^{-1/2} \; T^{5/4} \; \bar{m}^{-3/4} \; \exp\frac{-E_G}{4kT} \tag{4.35}$$

where $\bar{m}$ is the geometric mean effective mass, so that

$$\mathscr{E} \propto \kappa^{-1/2} \; E_g^{1/2} \; \bar{m}^{-3/4} \; T^{3/4} \tag{4.36}$$

In a more careful treatment, $\bar{m}$ would undoubtedly be replaced by the mass of the surface majority carrier. For germanium, field strengths in excess of 10^5 V/cm are predicted. As discussed previously (in Section 2.4 of Ref. [15]), and more recently by Greene [45], this is getting into the range where the underlying assumptions are questionable. Two new effects may be expected to enter the picture (cf. Section 3): (1) the boundary conditions will significantly affect the wave functions, and (2) the charge density will not depend only on the local potential. Both of these will be discussed in Section 5.

4.2.2. Level-Filling and Band Degeneracy

Still within the framework of thick samples and nonquantal conditions, we next consider the modifications needed when the Fermi level gets near or beyond a band edge. In a semiconductor with discrete defect levels, this implies also a change in occupancy of some of these defect levels. The theory was worked out by Seiwatz and Green [46] for the general case of degeneracy either in the bulk or at the surface or both. Since we have seen, however, that degeneracy in the bulk usually requires such high defect concentrations that the bands become distorted, we shall mainly be interested in degenerate surface layers on a nondegenerate bulk. Then only two of the terms in Eq. (4.22) are affected. Let us consider, for definiteness, an n-type semiconductor with a degenerate inversion layer. Then the previous approximations must be abandoned in Eqs. (4.13) and (4.15). As a result, Eq. (4.22a) becomes

$$|F(u, u_B)| = \left\{\int_u^{u_B}\left[-e^{u'} + \frac{\mathscr{F}_{1/2}(w_V - u')}{\mathscr{F}_{1/2}(w_V)}\right.\right.$$
$$\left.\left. + (e^{u_B} - e^{-u_B} + \left(\frac{N_A}{n_i}\right)\frac{1}{1 + \exp(u' - w_A')}\right]du'\right\}^{1/2} \tag{4.37}$$

for $u < 0 < u_B$, where, as a reminder, $w_V = \beta(E_V - E_i)$ and $w_A' = \beta(E_A' - E_i)$. The integral is still readily evaluated. We shall not repeat the results here, but only mention the physical effects. The second term is seen (by comparison of Eqs. (3.7) and (3.8), for example) to give a smaller hole charge than the nondegenerate approximation. Thus, the field strength is somewhat diminished and the layer widened. This can be seen clearly in the graphs presented by Many et al. (Figs. 4.7-4.9 of their text [15]). However, the last term, representing a positive charge contribution due to the neutralization of acceptors, tends to increase the field where the acceptor levels are near the Fermi level. The effect is limited, of course, since in this example the acceptors are the minority defects. The same effect could occur much more strongly in an accumulation layer. However, unless the defect levels are many kT away from the band edge, the field-weakening effect of the free-carrier degeneracy will still dominate.

Degeneracy in both bands is of interest mainly for semimetals. BenDaniel and Duke [47] worked out the theory for an intrinsic semimetal, and it is easy to extend this to the general case. With Eqs. (3.32) we obtain*

$$
|F(u, u_B)| = \left\{ \int_{u_B}^{u} \left[\left(1 + \frac{u'}{|w_C|}\right)^{3/2} - \left(1 - \frac{u'}{w_V}\right)^{3/2} \right. \right.
$$

$$
\left. \left. - \left(1 + \frac{u_B}{|w_C|}\right)^{3/2} + \left(1 - \frac{u_B}{w_V}\right)^{3/2} \right] du' \right\}^{1/2}
$$

*In Ref. [47], step-function factors are included to cut off the terms if the Fermi level crosses a band edge, but (a) it is unlikely that this can occur in the usual semimetals, and (b) if it did then the full Fermi-Dirac integral, not the degenerate approximation, would have to be used in the expressions for the charge density.

$$= \left\{ \frac{2}{5} |w_C| \left[\left(1 + \frac{u}{|w_C|}\right)^{5/2} - \left(1 + \frac{u_B}{|w_C|}\right)^{5/2} \right] \right.$$

$$+ \frac{2}{5} w_V \left[\left(1 - \frac{u}{w_V}\right)^{5/2} - \left(1 - \frac{u_B}{w_V}\right)^{5/2} \right]$$

$$\left. + (u_B - u) \left[\left(1 + \frac{u_B}{|w_C|}\right)^{3/2} - \left(1 - \frac{u_B}{w_V}\right)^{3/2} \right] \right\}^{1/2} \quad (4.38)$$

Since only small band bendings are to be expected, a linearized form of the theory is often an adequate approximation. On expanding the first four radicals in powers of $(u - u_B)$, we obtain

$$|F(u, u_b)| \simeq \left[\frac{3}{4} \left(\frac{|w_C| + u_B}{|w_C|^3} \right)^{1/2} + \frac{3}{4} \left(\frac{w_V - u_B}{w_V^3} \right)^{1/2} \right]^{1/2} |u - u_B| \quad (4.39)$$

Thus, in this approximation the barrier is exponential with a characteristic length L_B given by

$$\frac{1}{L_B^2} = \frac{3}{4L_D^2} \left[|\frac{1}{w_C}| \left(1 + \frac{u_B}{|w_C|}\right)^{1/2} + \frac{1}{w_V} \left(1 - \frac{u_B}{w_V}\right)^{1/2} \right] \quad (4.40)$$

This reduces to BenDaniel and Duke's expression when $u_B = 0$, and does not change much for modest values of $|u_B|$. In an illustrative example BenDaniel and Duke show that the linearized theory is indeed reasonably accurate.

4.3. METHOD OF SOLUTION FOR THIN SAMPLES

The thin-slab problem faces the great practical drawback that it is difficult to obtain the information needed for the boundary condition, Eq. (4.10). Indeed, one of the important applications of space charge theory is to measure Q_{int} from thick-slab experiments. Nevertheless, the thin-slab problem is certain to come more to the fore as the growing needs of thin-film technology demand it. Previous treatments have been rather scanty. We shall here extend the discussion (Section 2.3 of Ref. [16]) based on earlier work of Greene et al. [48] slightly, so as to be a bit more explicit about the procedure in a typical thin-slab problem.

Consider a slab with thickness ℓ and assume the surface field strengths $\mathscr{E}_z(z_0)$ and $\mathscr{E}_z(z_1)$ to be known. From Eq. (4.4) we obtain

$$
\mathscr{E}^2(z_1) - \mathscr{E}^2(z_0) = - \frac{2}{\kappa \varepsilon_0} \frac{kT}{e} \int_{u(z_0)}^{u(z_1)} \rho(u')du'
$$

$$
= \frac{2n_i kT}{\kappa \varepsilon_0} [F(u_1, u_B) - F(u_0, u_B)] \tag{4.41}
$$

This is an implicit relation between u_1 and u_0. It holds for any of the preceding forms of F. A second relation is obtained from the same equation written for $u(z)$; after solving for $|du/dz|$ and integrating, this gives

$$
z_1 - z_0 = \int_{u_0}^{u_1} \frac{(- \text{sgn } \mathscr{E}_z)du}{\left[\left(\frac{du}{dz}\right)^2_{z_0} + L_D^{-2} [F^2(u, u_B) - F^2(u_0, u_B)] \right]^{1/2}} \tag{4.42}
$$

A suitable computation procedure therefore is to guess a value of u_0, solve Eq. (4.41) for u_1, carry out the integration in Eq. (4.42) and see if the result equals ℓ, a somewhat unpalatable but not totally unfeasible procedure.

The solution for the completely symmetrical situation (equal values of u_S on the two faces) in the nondegenerate case has been given by Gasanov [49]. He also calculated the carrier excesses and the excess conductance (assuming bulk mobilities). One result is that inversion sets in earlier in a thin slab owing to the absence of the thick depletion region. However, the rather special condition of complete symmetry is probably not often encountered in practice. The quasi-symmetrical case had also been treated earlier by Albers and Thomas [50].

5. QUANTIZATION

It was seen in the preceding section that the macroscopic potential could vary by an appreciable amount in a very short distance near the surface of a conductor. Thus, carriers of one sign or the other will be constrained to move in a thin layer at the surface, and if the layer thickness becomes so small as to be comparable with the de Broglie wavelengths of the carriers, then quantization effects may be expected. This was first noted by Schrieffer [51] some fifteen years ago, and is now recognized as one of the general class of "size quantization" [52-56] effects due to confinement of particles in one or more dimensions by any means whatever.

Problems in which macroscopic fields are superimposed on a periodic potential are usually handled in the "effective mass approximation" of Slater [57] and Kohn [30]. The envelope function $\eta(\underline{r})$, which describes how the amplitude of the wave function $\psi(\underline{r})$ (for a state of a given band) varies from cell to cell in the crystal, is itself the solution of a Schrödinger-like equation. For our present one-dimensional problem, the latter is

$$\left(-\frac{\hbar^2}{2m_3} \frac{d^2}{dz^2} + qV - E_n \right) \eta_n(z) = 0 \tag{5.1}$$

where m_3 is the perpendicular effective mass. The generalization to the case where the effective mass tensor ellipsoid does not have a principal axis normal to the surface is given in Ref. [58]. Since periodic boundary conditions are appropriate in the x, y plane, ψ will have Bloch-like factors in these directions.

The justification for the effective mass equation is discussed by a number of authors, perhaps most completely by Weinreich [59]. The requirement is that V must vary only slowly with position. This means slowly enough so that its interband matrix elements are negligible. As has been suggested [16], it is not certain that this condition is met in strong space charge fields. Nevertheless, no other tractable theory is available.

5.1. BOUNDARY CONDITIONS

Clearly, some new boundary conditions must now be brought into the problem. In addition to those on the potential, Eqs. (4.5) and (4.9c), conditions on the wave function ψ are required. The exact condition is that ψ must be a smooth function which asymptotically approaches a form appropriate to each medium at large distances from the interface [60]. Unfortunately, this is not very useful in the present problem, and boundary conditions on the envelope function η are desired. Although, according to Fredkin and Wannier [61], all such conditions are ad hoc, most authors have followed Conley et al. [62] in assuming

(1) $\eta(z)$ continuous at z_S, and

(2) $m_3^{-1} \partial \eta / \partial Z$ continuous at z_S.

The second assumption is said to guarantee current continuity, but since the current density at any point is given [63] by $(\hbar/2mi)(\psi^* \operatorname{grad} \psi - \psi \operatorname{grad} \psi^*)$, it is not clear that this assumption correctly describes its average over the interface. However, this problem is more germane to tunneling than to quantization. For a junction between a conductor and a thick insulator, it is reasonable to take, in the conductor

$$\eta(z_S) = 0 \tag{5.2}$$

Indeed, Alferieff and Duke [64] show that the exact boundary conditions make very little difference once the potential step height exceeds a few tenths of an eV.

5.2. QUANTIZATION IN A MINORITY BAND

A special situation exists in the valence bands of germanium and silicon. Owing to the degeneracy at the band edge [65] there are "light" and "heavy" hole states and the former have both a lower density of states and a longer de Broglie wavelength. Thus, in a p-type surface layer the charge density is mainly given by the heavier carriers, and the states of the lighter ones

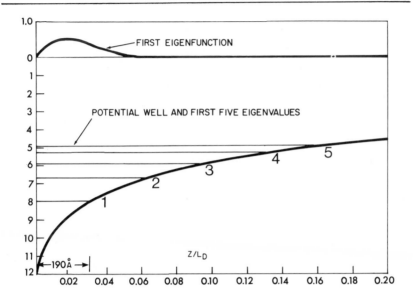

FIG. 5.1. Quantized light-hole energy levels and envelope function for the lowest level in germanium (after Handler and Eisenhour [66]).

are quantized in the resulting potential well. Since the well shape is known from the results of Section 4, it is a straightforward matter numerically to solve Eq. (5.1). This was done for germanium by Handler and Eisenhour [66] with the results shown in Fig. 5.1. As expected, the boundary condition keeps the carriers away from the surface.

5.3. QUANTIZATION IN A MAJORITY BAND

If the carriers contributing the major part of the space charge are quantized, then the potential to be used in Eq. (5.1) is no longer known a priori. A self-consistent calculation should be carried through, but since this is always a difficult matter, several authors have worked with assumed forms for the potential, based usually on the nonquantal space charge theory. An early example is the work of BenDaniel and Duke [47] who took, for a semimetal,

$$qV(z) = - eV_S \exp (- z/L_B) \tag{5.3}$$

Then Eq. (5.1) is easily reduced to the Bessel equation so that the bound-state envelope functions are [67]

$$\eta_n(z) = J_{\nu_n}(y) \tag{5.4}$$

where

$$\nu_n = 2L_B [- 2m_3 E_n/\hbar^2]^{1/2} \tag{5.5}$$

and

$$y = 2L_B (2m_3 eV_S/\hbar^2)^{1/2} \exp (- z/2L_B) \tag{5.6}$$

Each eigenvalue E_n leads to a uniform surface density of states

$$N_n = g_n (m_1 m_2)^{1/2}/2\pi\hbar^2 \tag{5.7}$$

where m_1 and m_2 are the effective masses parallel to the surface and g_n is the degeneracy of the level E_n. There is, of course, one such state for each ellipsoid in the band under consideration, so all the above quantities should be labelled with still another index i. Thus, the total charge density in these states will be

$$\rho(z) = q \sum_{n, i} N_n^{(i)} |\eta_n^{(i)}(z)|^2 (E_F - E_n^{(i)}) S(E_F - E_n^{(i)}) \tag{5.8}$$

where S(x) is the unit step function.

In general, $\rho(z)$ is only part of the space charge; the contributions of the ionized defects and the repelled carriers must also be included. However, in a semiconductor, under conditions where quantization will be important, Eq. (5.8) gives essentially the entire charge density in the vicinity of the surface. Hence $\rho(z)$ can be differentiated twice to find a corrected potential for use in the next cycle of a self-consistent calculation. Alferieff and Duke [64] show that the corrected potential is considerably steeper than the starting form. Duke [67] made a "semi-self-consistent" calculation by

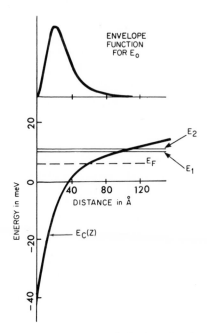

FIG. 5.2. Self-consistent quantized electron levels and envelope function for the lowest level. (100) surface of p-type silicon (75 ohm-cm bulk resistivity). Fermi level shown for 10^{12} electrons/cm^2 in the two dimensional band based on the lowest level, E_0. (Adapted from Stern and Howard [58].)

retaining the exponential form, Eq. (5.3), but adjusting the parameters V_S and L_B until the total trapped charge [integral of $\rho(z)dz$] was correct.

A fully self-consistent calculation for the conduction bands of InAs and Si is reported by Stern and Howard [58] who give as an approximate expression

$$|\eta(z)|^2 \simeq \frac{1}{2} b^3 z^2 \exp(-bz) \tag{5.9}$$

An example of the energy spectrum they obtained is shown in Fig. 5.2.

5.4. SURFACE LANDAU LEVELS

Although quantization exerts a number of subtle influences on the transport properties of space charge layers [45], is shows up most dramatically when present in conjunction with a strong magnetic field perpendicular to the surface. The reason is that, since motion in the z direction is already quantized, there is no continuum of energies corresponding to momenta along the field. Thus, the Landau levels [68] are truly discrete energy levels rather than merely edges of sub-bands, and for this reason very sharp changes in physical properties can occur as the Fermi level is moved through them. The first example was seen in the conductance of silicon [69]. Related effects in capacitance [70] will be discussed in Section 6.5.

6. MIS STRUCTURES

The abbreviation "MIS" stands for metal-insulator-semiconductor (or semimetal) and thus describes a sandwich configuration to which a voltage can be applied. (The possibly more familiar "MOS" is the same except that the insulator is specifically designated as an oxide, usually of the semiconductor). Through a discussion of MIS structures, one can readily understand a variety of other structures, such as the Schottky barrier [3] or MS junction, which is obtained from the MIS by letting the insulator thickness go to zero, as well as the p-n junction (see Ref. [14], p. 86, where references to the literature are given) and the heterojunction [71], which result if the metal in a Schottky barrier is replaced by a semiconductor.

Figure 6.1 illustrates the relations among the various parameters for an MIS with voltage V_M applied to the metal relative to the bulk of the semiconductor. From the figure, it is seen that

$$V_M = V_I + V_S + V_0^{(M, S)} \qquad (6.1)$$

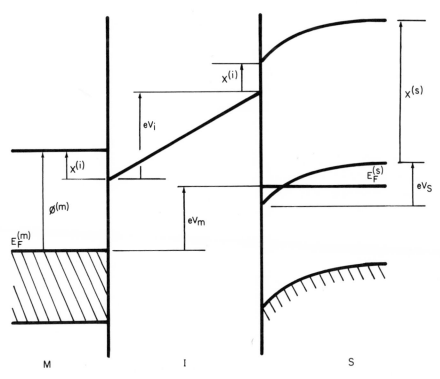

FIG. 6.1. Energy levels in the MIS structure with voltage V_M applied to the metal relative to the semiconductor bulk.

where

$$eV_M \equiv E_F^{(S)} - E_F^{(M)} \tag{6.2}$$

and

$$eV_0^{(M, S)} = \phi^{(M)} - \chi^{(S)} - (E_{CB} - E_F)^{(S)} \tag{6.3}$$

$$= \text{a constant, characteristic of the two conductive materials}$$

If there is no volume charge in the dielectric (as is assumed in Fig. 6.1) then the field there is uniform, and from Eqs. (4.10) and (4.29a) we obtain

$$\kappa_I \mathscr{E}_z^{(I)} = -(Q_{SC} + Q_{int})/\varepsilon_0 \tag{6.4}$$

or

$$V_I = -(C_I)^{-1} (Q_{SC} + Q_{int}) \tag{6.5}$$

where

$$C_I = \kappa_I \varepsilon_0 / \ell \tag{6.6}$$

= capacitance of the dielectric layer per unit area

and where Q_{int} is given in Eq. (4.10a). It will be noted that the charges on the right side of Eq. (6.5) are both functions of V_S.

6.1. SURFACE CAPACITANCE

In recent years, the development of technologies [72] for the preparation of thin high-quality insulating layers has made it possible to achieve very large values of C_I. Thus, V_I is relatively small so that an appreciable part of V_M appears across the space charge layer. This has greatly enhanced the utility of the surface capacitance measurement as a tool for studying a variety of properties of the surface and its vicinity, as we shall now discuss.

The applied voltage is varied with time as

$$V_M(t) = \bar{V}_M + \tilde{V}_M \cos \omega t \tag{6.7}$$

where $\bar{V}_M$ is the "steady" (i.e., slowly varying) "gate-bias" voltage and $\tilde{V}_M$ is the small "measuring" voltage. The MIS response is purely capacitive provided $\tilde{Q}_M$ is in phase with $\tilde{V}_M$. This can of course occur only in certain frequency ranges because of the finite rates of change of the various charge contributions. But in these ranges we have

$$C \equiv dQ_M/dV_M = - d(Q_{SC} + Q_{int})/dV_M \tag{6.8}$$

Thus, from Eq. (6.1),

$$C^{-1} = (C_I)^{-1} + (C_{SS} + C_{SC})^{-1} \tag{6.9}$$

where

$$C_{SS} = - dQ_{SS}/dV_S \qquad (6.10)$$

and

$$C_{SC} = - dQ_{SC}/dV_S \qquad (6.11a)$$

$$= - \frac{\epsilon_0}{L_D} \frac{\partial}{\partial u_S} F(u_S, u_B) \qquad (6.11b)$$

Further, since Q_{SC} is in general the sum of charge contributions due to majority carriers, minority carriers, and defects, C_{SC} may be written as the sum of three terms

$$C_{SC} = C_{maj} + C_{min} + C_{def} \qquad (6.12)$$

Each of these will have its own characteristic relaxation time τ, so that

$$C_{(\ )}(\omega, \overline{V}_S) = C_{(\ )}(0, \overline{V}_S) [1 + i\omega\tau_{(\ )}(\overline{V}_S)]^{-1} \qquad (6.13)$$

Here, the notation emphasizes the fact that the low-frequency capacitances and their relaxation times are functions of the steady-state (average) surface potential. The surface state capacitance may be written as a sum of similar terms, one for each type of surface state present. In practice, the surface state relaxation times may vary over a tremendously wide range, from minutes or hours for slow states to fractions of a microsecond for fast states [73].

Because of the multiplicity of charge storage and transport mechanisms, and because of the three-dimensional nature of the space charge region, it is not really possible to exhibit any exact equivalent circuit for the MIS regarded as a two-terminal network. However, if the circuit elements are permitted to vary with both frequency and voltage, then various approximate equivalent circuits may be devised. One such, which may be helpful in our discussion, is shown in Fig. 6.2. Several rather more elaborate and comprehensive circuits have been suggested by Lehovec and Slobodskoy [74] among others.

6.2. DETERMINATION OF V_S

One of the important applications of capacitance measurements is the determination of V_S as function of V_M. This can be done only in certain

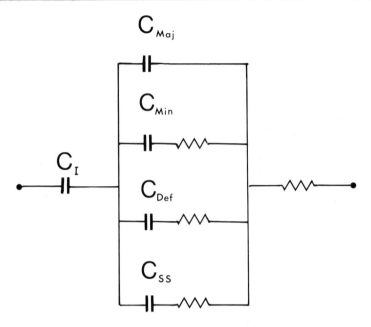

FIG. 6.2. An approximate equivalent circuit for an MIS structure.

favorable cases, namely when $C_{SS}(\omega)$ is either (1) equal to $C_{SS}(0)$ or (2) negligible. Case (1) pertains when ω can be made so small that $\omega\tau \ll 1$ for all the pertinent surface states. This can occur when slow states are absent, and also when they are present but so very slow that they absorb no charge during the entire period that V_M is applied. Then since

$$dV_S = dQ_M/(C_{SS} + C_{SC})$$
$$= (C^{-1} - C_I^{-1})C \, dV_M \tag{6.14}$$

we find

$$V_S(V_M) = V_S(V_{ref}) + \int_{V_{ref}}^{V_M} \left[1 - \frac{C(V_M')}{C_I} \right] dV_M' \tag{6.15}$$

A convenient reference voltage is the flat–band voltage V_{FB}, which is the value of V_M for which $V_S = 0$. Thus, alternatively,

$$V_S(V_M) = \int_{V_{FB}}^{V_M} [1 - C/C_I]dV_M' \qquad (6.16)$$

It will be noted that V_S is thus obtained (within an additive constant) directly from the measurements.

Case (2) pertains when the fast states are either absent or not too fast, so that it is possible to make $\omega\tau \gg 1$ for all states. Then

$$(C^{-1} - C_I^{-1})_{hf} = C_{SC}^{-1} \qquad (6.17)$$

and C_{SC} is a theoretically known function of V_S under suitable conditions. These conditions are that $\omega\tau$ must be either $\gg 1$ or $\ll 1$ for each of the three contributions to C_{SS} in Eq. (6.12). For C_{maj}, there is usually no problem: since only the motion of carriers in the majority band is involved, the relaxation time is very short. For C_{min}, several possibilities exist. When $\overline{V}_S$ is such that an accumulation or moderate depletion layer is present, not only is C_{min} quite small but the mechanism is again carrier motion, so that $\omega\tau$ is again $\ll 1$. For an inversion layer, however, the large concentrations of minority carriers needed are not available from the bulk and must usually enter and leave the surface layer by generation and recombination processes [75-77]. Such processes can range from very rapid to very slow, depending on the materials and the details of the transition mechanisms.

Case 1. Inversion Layer in Equilibrium

If transitions are sufficiently rapid, the minority carriers remain fully in equilibrium. Leaving aside for the moment any questions of trapping at defects, C_{SC} is then obtained from Eq. (6.11) using the equilibrium expressions for Q_{SC} discussed in Sections 4 and 5. For the simple case depicted in Fig. 4.4, C_{SC} clearly is very large in both accumulation and strong inversion layers, with a minimum in the mild inversion range. The exact position of the minimum is given [78] by the solution of the equation

$$\cosh u_S = [\sinh u_S - \sinh u_B/F(u_S, u_B)]^2 \qquad (6.18)$$

Some examples of the behavior of $C(V_M)$ are also given in the reference just cited.

The capacitance of thin slabs in the completely symmetrical case has been treated by Gasanov and Stafeev [79]. While this rather idealized case is not likely to be encountered in practice, the general result that the minimum capacitance is lowered may be of use for some device applications, such as low-frequency varactors.

Case 2. Inversion Layer Out of Equilibrium

If the generation and recombination processes are slow, the minority carrier charge in the inversion layer cannot (with some exceptions to be discussed later) follow the high-frequency signal. A special sub-case occurs if the layer cannot even follow the bias voltage $\overline{V}_M$. Then no inversion layer forms, and the parabolic potential region (the depletion layer) continues to widen as $|\overline{V}_M|$ is increased in the inversion direction. This case can be treated by simply omitting the minority carrier terms in the expression for $|F(u, u_B)|$, Eq. (4.22c). The result is that the capacitance continues to decrease beyond the equilibrium minimum. The approximation neglects the fact that some of the minority carriers (those near the bulk) can always follow the voltage to some extent, so the calculated decrease is a bit too rapid.

The more common and important situation is that in which the minority carriers can follow $\overline{V}_M$ but not $\tilde{V}_M$, so that a well developed inversion layer is present but is out of equilibrium with the bulk. Indeed, the difference of electrochemical potential between the two regions can be made very large by connecting a battery between the bulk and a contact to the inversion layer (i.e., a region that intersects the surface and is doped oppositely to the bulk). The theory of such channels was first developed by Statz et al. [80] and is summarized in the texts. The more recent adaptation to the capacitance has been made mainly by Grove and co-workers [81-83] and, for the case of the unbiased channel, by Lehovec et al. [84], and Berz [85] based on the high-frequency field effect theory of Garrett [86], Berz [87], and Yunovich [88].

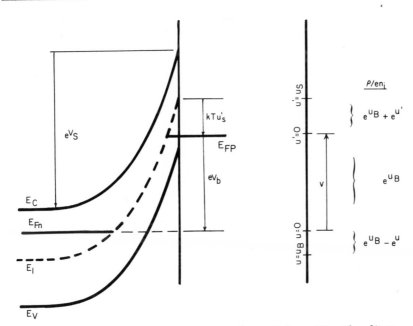

FIG. 6.3. Energy levels in an inversion layer ("channel") with voltage V_b applied relative to the bulk. Scale of u and u', and dominant charge contributions shown at right.

Since the system is not in equilibrium there is no single Fermi energy at any point. However, it is an excellent approximation [89] to take the quasi-Fermi levels (see Ref. [14], pp. 308ff) for the carriers that are locally in the majority as constant in each of the highly conductive regions. The situation for an n-type bulk is depicted in Fig. 6.3. There is now another variable (V_b or its dimensionless equivalent, $v = eV_b/kT$ in the problem, and Eqs. (4.29c) and (4.30) are generalized to read

$$Q_{SC} = 2en_i L_D \ F(u_S', u_B, v) \qquad (6.19)$$

$$\Delta P = n_i L_D \ G(u_S', u_B, v) \qquad (6.20)$$

In the approximation of retaining, in each region, only the main contribution to the charge density, the functions are [80]

$$
\begin{aligned}
|F(u_S', u_B, v)| &= \left\{ \int_{u_B}^0 (-e^{u_B} + e^u)du \right. \\
&\left. + \int_0^v - e^{u_B}du + \int_0^{u_S'} (-e^{u_B} - e^{-u'})du' \right\}^{1/2} \\
&= \left\{ (u_B - v - u_S' - 1)e^{u_B} + e^{-u_S'} \right\}^{1/2}
\end{aligned} \tag{6.21}
$$

and

$$
G(u_S', u_B, v) = \int_0^{u_S'} \frac{e^{-u'} du'}{F(u', u_B, v)} \tag{6.22}
$$

For illustrative purposes, conventional plots of the two functions are shown in Fig. 6.4. It should be noted that $G(u_S', u_B, 0)$ is not exactly the same as the equilibrium function $G(u_S, u_B)$, since we here are counting the minority carriers only in the inversion region. Another mode of presentation of the data has been introduced by Grove and co-workers [81-83]. They plot G vs F for constant values of v. Lines of constant u_S [82] or u_S' [83] may also be included.

Now returning to the capacitance, there are, once again, two cases to be considered. If a definite substrate-to-channel bias V_b is imposed, then clearly v is held constant and

$$
C_{SC}^{(v)} = - \frac{\kappa\varepsilon_0}{L_D} \left(\frac{\partial F}{\partial u_S'} \right)_{v = const.} \tag{6.23}
$$

The value can be read off a graph such as Fig. 6.4(a), or readily calculated from Eq. (6.21). Some typical results are shown by Grove and Fitzgerald [83]. It will be noted that these resemble the low-frequency case in that C_{SC} becomes very large at sufficiently large values of u_S'. The reason is that the biasing source which serves to hold v constant is, in effect, a reservoir of minority carriers. A similar reservoir effect (i.e., high inversion capacitance even at high frequencies) is seen even without a biasing source if there is a strong inversion layer in the area surrounding the active region of the surface [90].

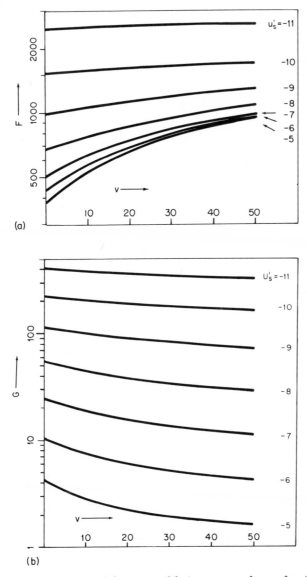

FIG. 6.4. Illustration of the nonequilibrium space charge functions:
(a) $F(u_s', u_B, v)$ and (b) $G(u_s', u_B, v)$ defined in Eqs. (6.21) and (6.22), respectively. Values shown are for $u_B = 5$.

In the more usual case, there is no definite bias, and no other reservoir, so that the total number of carriers in the channel is fixed (with respect to $\tilde{V}_M$). Thus

$$C_{SC}^{(G)} = -\frac{\kappa\varepsilon_0}{L_D}\left(\frac{\partial F}{\partial u_S}\right)_{G\,=\,const.} \tag{6.24a}$$

$$= -\frac{\kappa\varepsilon_0}{L_D}\left[\frac{\partial F}{\partial u_S} - \left(\frac{\partial F}{\partial v}\right)\left(\frac{\partial G}{\partial u_S}\right)\left(\frac{\partial G}{\partial v}\right)^{-1}\right] \tag{6.24b}$$

It is seen from Fig. 6.2 that this is much smaller than the right side of Eq. (6.23). Grove et al. [81] present a nice intuitive picture of $C_{SC}^{(G)}$ as simply the capacitance of the depleted layer of the semiconductor. The width of this layer becomes essentially constant once an inversion layer forms.

To sum up the foregoing rather lengthy discussion, the surface potential may be determined as function of $\overline{V}_M$ with good accuracy if:

(1) there are no surface states with time constants between the periods of the measuring voltage $\tilde{V}_M$ and the gate bias voltage $\overline{V}_M$; then Eq. (6.16) applies; or

(2) there are no surface states faster than the period of the measuring voltage, and (a) the minority carrier generation and recombination processes are faster than this, or (b) a reservoir of minority carriers is present. If conditions (a) and (b) are not met, the inversion-layer capacitance is very insensitive to V_S. Lehovec et al. [84] have attempted to utilize the slight variation of this capacitance, but the results are probably not too reliable.

6.3. SURFACE STATES (AND PSEUDO-SURFACE STATES)

Investigation of the parameters of surface states is one of the obvious attractive applications of capacitance measurements. It has long been recognized [91] that the capacitance method is in principle superior to the classical field-effect conductance method, since no knowledge of surface mobilities [45] is required. However, a variety of technical difficulties, not the least of which is the requirement that the series capacitance be very large, have so far restricted its applicability.

Two general approaches to the surface state problem are possible. To obtain the total density of states of all response times, one simply compares the measured total induced charge [cf. Eq. (6.8)]

$$- Q_M(V_M) = - \int_0^{\overline{V}_M} C(\overline{V}_M{}')dV_M{}' \tag{6.25}$$

with the calculated space charge

$$Q_{SC}(\overline{V}_M) = Q_{SC}(V_S[\overline{V}_M]) \tag{6.26}$$

The difference, taken as function of V_S, yields the desired information. This is the approach taken by Lehovec et al. [84], Terman [92], and many others. Gray and Brown [93] found a clever way to extend the range of V_S values by varying the temperature. Their results for oxidized silicon suggested high concentrations of states in narrow energy ranges about 0.1 eV from the band edges.

As for information about the time constants of the faster states, it is clear that this is contained in the dispersion (frequency dependence) of the capacitance, but what is not at all clear is how a quantitative interpretation is to be made. Some typical results are cited by Terman [92]. Zaininger and Warfield [94] give a detailed critique of the interpretational difficulties and conclude that very little fundamental information is to be gleaned. The difficulty is somewhat alleviated by the measurement of the real part of the admittance, as suggested by Nicollian and Goetzberger [95] but the fundamental problems remain.

One of the problems is a result of the fact that the measured trapped charge includes all charges in the vicinity of the surface. Some of the charges, for example, may be held in defects in the insulator [96] at various distances from the interface. In this case the measured time constants would relate to the distance rather than to any property of the trap state. Another possibility is that some charges could be drawn to the surface by the attraction of fixed charges in the insulator [97]. All such situations in which trapping is due to the insulator, not the semiconductor surface itself, should probably be designated as pseudo-surface states. Measurements thereon may be of great practical importance from the standpoint of device operation [98] but can hardly be expected to shed much light on fundamental surface properties. There exists by now a very extensive literature relating such results to the details of preparation and treatment of the dielectric layer but we shall not attempt to review it here.

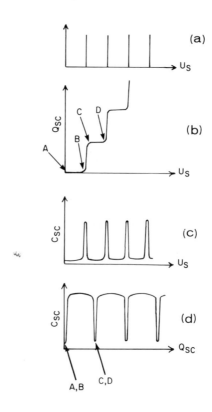

FIG. 6.5. Surface Landau levels: (a) density of states; (b) trapped charge; (c) capacitance as function of surface potential; (d) capacitance as function of trapped charge (i.e., essentially as function of applied voltage). A, B, C, ··· label corresponding points in parts (b) and (d).

6.4. SURFACE QUANTIZATION EFFECTS

A most interesting application of MIS capacitance measurements has been made by Kaplit and Zemel [70] to give a dramatic demonstration of the quantization of levels in a narrow space charge layer. A strong magnetic field was applied normal to the surface so that the two-dimensional bands were split into Landau levels. Thus, as discussed in Section 5, there is a series of highly degenerate discrete states. The basis of the experiments is best discussed with reference to Fig. 6.5. Part (a) shows the density of states, a series of delta-functions. As u_S is varied the electron concentration thus follows a series of Fermi-Dirac functions, as in Fig. 6.5(b), and the resulting space charge capacitance is proportional to the slope of these functions, as in Fig. 6.5(c). Now, the field strength in the dielectric layer is proportional to Q_{SC} by Eqs. (4.10) and (4.29a), if there is no appreciable concentration of surface states. Thus $\overline{V}_M$ is essentially proportional to Q_{SC}. Accordingly, in Fig. 6.5(d), C_{SC} is plotted against Q_{SC}. Thus, it is clear that C vs $\overline{V}_M$ will comprise a series of fairly sharp dips below the background level C_I. It is seen that the sharp dips do not correspond to the Landau levels, but rather to the broad state-free valleys between them. Nevertheless, the results provide unambiguous confirmation of the discreteness of the levels.

7. TRAPPED CHARGE DISTRIBUTION

As was seen in Section 6.3, it is often possible for charges to be trapped within the volume of the insulating layer. Such charges may have some mobility under special treatment conditions (e.g., high temperature) but are essentially fixed under the usual conditions of measurement. Hence, their presence does not affect the value of C_I or the relation of C_S to V_S, but does affect the value of V_I in Eq. (6.1). Also, in Eq. (6.4) $\mathscr{E}_z^{(I)}$ obviously must be taken as $\mathscr{E}_z^{(I)}(0)$, the field strength in the dielectric at the interface.

If the charge distribution $\rho^{(I)}(z)$ is known, Poisson's equation may be integrated twice to obtain [99, 100]

$$V^{(I)}(z) = V_S - z\,\mathscr{E}_z^{(I)}(0) - \frac{1}{\kappa_I \epsilon_0}\int_0^z dz' \int_0^{z'} \rho(z'')dz'' \tag{7.1}$$

Thus,

$$V_I \equiv V^{(I)}(-\ell) - V_S \tag{7.2a}$$

$$= -(C_I)^{-1}\left[Q_{SC} + Q_{int} + \frac{1}{\ell}\int_0^{-\ell} dz \int_0^z \rho(\dot{z}')dz'\right] \tag{7.2b}$$

$$= -(C_I)^{-1}\left[Q_{SC} + Q_{int} - Q_I - \frac{1}{\ell}\int_0^{-\ell} z\rho(z)dz\right] \tag{7.2c}$$

$$= -(C_I)^{-1}\left[Q_{SC} + Q_{int} - \frac{1}{\ell}\int_0^{-\ell} (z+\ell)\rho(z)dz\right] \tag{7.2d}$$

It is seen that smaller values of $|z|$ increase the magnitude of the last term in Eq. (7.2d). In other words, a given amount of charge in the insulator has its maximum effect on V_I when it is located as close as possible to the semiconductor, and has no effect when located at the metal.

This raises the question of how a fixed amount of slightly mobile charge will distribute itself in the insulator. In many cases (i.e., in the absence of complications due to chemical reactions) this can be decided on the basis of minimization of the total electrostatic energy of the system [100]. The result turns out to depend on several parameters, but in thick dielectric layers there is a tendency for charges to move to the interface. This will not only produce the maximum shift in the $V_I(V_S)$ relation, but will also tend to introduce pseudo-surface states (cf. Section 6.3). It is for these reasons that tremendous efforts have been made to produce ultrapure dielectric layers for device applications.

We now turn to a consideration of the distribution of charge in the two dimensions parallel to the surface. There are two questions to be considered: (1) how grainy is the actual charge distribution, and (2) how seriously does this vitiate the one-dimensional type of theory that we have used throughout this chapter? With regard to the first question, the answer obviously depends on the nature of the charges constituting Q_{int}. For example, electrons (or holes) trapped in intrinsic surface states [101, 102] will be quite uniformly distributed over the interface plane. Pseudo-surface-state wave functions [97] are somewhat more concentrated, while adsorbed or trapped ions may be very concentrated. Of course, even in the latter case a small net charge density that results from the near cancellation of large positive and negative densities may be quite uniform. Thus, there are a number of situations in which the one-dimensional model should be reasonably good.

On the other hand, it now appears that when the surface charge distribution is highly discrete, the one-dimensional approximation fails rather badly. Greene and co-workers [103, 104] have investigated a variety of problems involving point charges at interfaces. The result that concerns us here is that the one-dimensional treatment does not, in general, correctly give even the average (on a plane) of the correct three-dimensional potential. The error was found to be serious in an accumulation layer under the one-carrier model investigated. Undoubtedly the same would be true in an inversion layer in a two-carrier model.

LIST OF SYMBOLS

C	Capacitance per unit area (F/m^2)
$\underline{D}$	Dielectric displacement (C/m^2)
e	Magnitude of electron charge $(+1.6 \times 10^{-19}$ C)
E_F	Fermi energy
E_i	Intrinsic Fermi energy
E_j	Energy of jth energy level
E_{vac}	Vacuum energy of an electron [Eq. (4.6)]
E^*	Energy level that equals E_{vac} at surface (cf. Fig. 4.1)
E'	Effective energy level of a defect [Eq. (3.12)]
$\mathscr{E}$	Electric field strength (V/m)
f_j	Occupation probability of jth energy level
F	Space charge function, equal to normalized field strength (Section 4)
$\mathscr{F}_j$	Fermi-Dirac integral [Eq. (3.6)]

g_s	Degeneracy of ground state of a defect in charge state s
G	Space charge function giving carrier excesses (Section 4)
$\underline{k}$	Wave vector (m^{-1})
L_B	Bulk screening length [Eq. (4.28)]
L_D	Intrinsic screening length (Debye length) [Eq. (4.27)]
m_C, m_V	Density-of-states effective mass in conduction and valence bands, respectively
n	Electron concentration
n_i	Intrinsic carrier concentration
N_C, N_V	Effective concentration of states in conduction and valence bands, respectively
N_s	Concentration of defects in charge state s
p	Hole concentration
$\underline{P}$	Polarization (C/m^2)
q	Charge of a particle
Q	Charge per unit area
$\overset{\leftrightarrow}{Q}$	Quadrupole moment density (C/m)
$\underline{r}$	Position vector
T	Temperature $(^\circ K)$
u	$\beta(E_F - E_i)$
U	Periodic factor in a Bloch wave function
V	Electrostatic potential (V)
$w_{(\ )}$	$\beta(E_{(\ )} - E_i)$
z	Distance into a material normal to surface or interface
Z_s	Partition function of a defect in charge state s
β	$(kT)^{-1}$ (k = Boltzmann constant)
$\Delta N, \Delta P$	Excess electron or hole concentration (per unit area) (m^{-2})
ε_0	Permittivity of space $(8.85 \times 10^{-12}\ F/m)$
η	Envelope factor in effective mass wave function [Eq. (5.7)]
κ	Relative permittivity (dielectric constant)
ρ	Charge density (C/m^3)
τ	Time constant (sec)
ϕ	Work function [Eq. (4.7)]
χ	Electron affinity [Eq. (5.1)]
ψ	Wave function
ω	Angular frequency (sec^{-1})

Subscripts and Superscripts

A	Acceptor
B	Bulk
C	Conduction band
D	Donor
def	Defect
I	Insulating layer
int	Interface
maj	Majority carrier
mic	Microscopic
min	Minority carrier
M	Metal (i.e., electrode, field plate, or gate of an MIS structure)
S	Surface (or interface)
s	Charge state of a defect
V	Valence band

REFERENCES

[1]   J. R. Macdonald, Solid State Electron, 5, 11 (1962)

[2]   N. F. Mott, Proc. Roy. Soc., Ser. A., 171, 27 (1939).

[3]   W. Schottky, Z. Phys., 113, 367 (1939).

[4]   N. F. Mott and R. W. Gurney, Electronic Processes in Ionic Crys-
       tals, 2nd ed., Oxford Univ. Press, Oxford, England, 1948.

[5]   H. K. Henisch, Rectifying Semiconductor Contacts, Oxford Univ.
       Press, Oxford, England, 1957.

[6]   E. Spenke, Electronic Semiconductors, McGraw-Hill, New York,
       1958.

[7]   P. Aigrain and C. Dugas, Z. Elektrochem., 56, 363 (1952).

[8]   K. Hauffe and H. J. Engell, Z. Elektrochem., 56, 366 (1952).

[9]   P. B. Weisz, J. Chem. Phys., 21, 1531 (1953).

[10]   W. H. Brattain and J. Bardeen, Bell Syst. Tech. J., 32, 1 (1953).

[11]   W. Shockley and G. L. Pearson, Phys. Rev., 74, 232 (1948).

[12] C. G. B. Garrett and W. H. Brattain, Phys. Rev., 99, 376 (1955).

[13] R. H. Kingston and S. F. Neustadter, J. Appl. Phys., 26, 718 (1955).

[14] W. Shockley, Electrons and Holes in Semiconductors, Van Nostrand, Princeton, N. J. (1950), p. 308.

[15] A. Many, Y. Goldstein, and N. B. Grover, Semiconductor Surfaces, North-Holland, Amsterdam, 1965.

[16] D. R. Frankl, Electrical Properties of Semiconductor Surfaces, Pergamon, London, 1967.

[17] J. D. Jackson, Classical Electrodynamics, Wiley, New York, 1962.

[18] G. Russakoff, Am. J. Phys., 38, 1188 (1970).

[19] H. Juretschke, Phys. Rev., 92, 1140 (1953).

[20] T. L. Loucks and P. H. Cutler, J. Phys. Chem. Solids, 25, 105 (1964).

[21] J. Bardeen, Surface Sci., 2, 381 (1964).

[22] A. Rabinovitch and J. Zak, Phys. Rev., B4, 2358 (1971).

[23] J. C. Slater, Phys. Rev., 87, 807 (1952).

[24] J. D. Levine, Phys. Rev., 171, 701 (1968).

[25] J. S. Blakemore, Semiconductor Statistics, Pergamon, London, 1962.

[26] A. H. Wilson, Theory of Metals, 2nd ed., Cambridge Univ. Press, Cambridge, England (1953), Appendix A.

[27] R. N. Hill, Amer. J. Phys., 38, 1440 (1970).

[28] W. Shockley and J. T. Last, Phys. Rev., 107, 392 (1957).

[29] J. M. Luttinger and W. Kohn, Phys. Rev., 97, 869 (1955).

[30] W. Kohn, in Solid State Physics, Vol. 5 (F. Seitz and D. Turnbull, eds.), Academic Press, New York, 1957, p. 258.

[31] J. Blakemore, Phil. Mag., 4, 560 (1959).

[32] J. I. Pankove, Progress in Semiconductors, Vol. 9 (A. F. Gibson and R. E. Burgess, eds.), Wiley, New York, 1965, p. 47.

[33] E. M. Conwell and B. W. Levinger, Proc. Internl. Conf. Phys. Semicond., Exeter (1962), Inst. of Physics and Physical Society, London, 1962, page 227.

[34] D. M. Kaplan, Bull. Amer. Phys. Soc., Ser. II, 16, 437 (1971).

[35] W. S. Boyle and G. E. Smith, in Progress in Semiconductors, Vol. 7 (A. F. Gibson and R. E. Burgess, eds.), Wiley, New York, 1963.

[36]   J. C. Rivière, in Solid State Surface Science, Vol. 1 (M. Green, ed.),
       Marcel Dekker, New York, 1969, p. 179.

[37]   D. V. Geppert, A. M. Cowley, and B. V. Dore, J. Appl. Phys., 37,
       2458 (1966).

[38]   D. R. Frankl, J. Appl. Phys., 31, 1752 (1960).

[39]   C. Goldberg, Solid State Electron., 7, 593 (1964).

[40]   H. Flietner, Ann. Phys. (Liepzig), 7, 396 (1959).

[41]   V. J. Lee and D. R. Mason, J. Appl. Phys., 34, 2660 (1963).

[42]   F. Berz, J. Phys. Chem. Solids, 25, 859 (1964).

[43]   G. C. Dousmanis and R. C. Duncan, Jr., J. Appl. Phys., 29, 1627
       (1958).

[44]   C. E. Young, J. Appl. Phys., 32, 329 (1961).

[45]   R. F. Greene, in Solid State Surface Science, Vol. 1 (M. Green, ed.),
       Marcel Dekker, New York, 1969, p. 87, Eq. (9.1).

[46]   R. Seiwatz and M. Green, J. Appl. Phys., 29, 1034 (1958).

[47]   D. J. BenDaniel and C. B. Duke, Phys. Rev., 152, 683 (1966).

[48]   R. F. Greene, D. R. Frankl, and J. Zemel, Phys. Rev., 118, 967
       (1960).

[49]   L. S. Gasanov, Soviet Phys.—Semicond., 1, 673 (1967).

[50]   W. A. Albers, Jr. and J. E. Thomas, Jr., Technical Note No. 2,
       U. S. Air Force Contract AF 49(638)-158 (March 1, 1960).

[51]   J. R. Schrieffer, Phys. Rev., 97, 641 (1955).

[52]   E. C. Crittenden, Jr. and K. W. Hoffman, Rev. Mod. Phys., 25, 310
       (1953).

[53]   I. M. Lifshitz and M. I. Kaganov, Soviet Phys.—Uspekhi, 2, 831
       (1960).

[54]   V. Bezak, J. Phys. Chem. Solids, 27, 815 (1966); J. Phys. Chem.
       Solids, 27, 821 (1966).

[55]   V. B. Sandomirskii, Soviet Phys.—JETP, 25, 101 (1967).

[56]   R. C. Jaklevic, J. Lambe, M. Mikkor, and W. C. Vassel, Phys.
       Rev. Lett., 26, 88 (1971).

[57]   J. C. Slater, Phys. Rev., 76, 1592 (1949).

[58]   F. Stern and W. E. Howard, Phys. Rev., 163, 816 (1967).

[59]   G. Weinreich, Solids: Elementary Theory for Advanced Students,
       Wiley, New York, 1965.

[60]   T. E. Feuchtwang, Phys. Rev., B, 2, 1863 (1970).

[61]   D. R. Fredkin and G. H. Wannier, Phys. Rev., 128, 2054 (1962).

[62]   J. W. Conley, C. B. Duke, G. D. Mahan, and J. J. Tiemann, Phys. Rev., 150, 466 (1966).

[63]   K. Gottfried, Quantum Mechanics, Benjamin, New York, 1966, page 20. (Also, many other texts on quantum mechanics.)

[64]   M. E. Alferieff and C. B. Duke, Phys. Rev., 168, 832 (1968).

[65]   J. Callaway, Energy Band Theory, Academic Press, New York, 1964, pp. 157ff.

[66]   P. Handler and S. Eisenhour, Surface Sci., 2, 64 (1964).

[67]   C. B. Duke, Phys. Rev., 159, 632 (1967).

[68]   J. M. Ziman, Electrons and Phonons, Oxford Univ. Press, Oxford, England, 1960, pp. 521-523.

[69]   A. B. Fowler, F. F. Fang, W. E. Howard, and P. J. Stiles, Phys. Rev. Lett., 16, 901 (1966).

[70]   M. Kaplit and J. N. Zemel, Phys. Rev. Lett., 21, 212 (1968).

[71]   R. L. Anderson, Solid State Electron., 5, 341 (1962).

[72]   Conference on Thin Film Dielectrics, Montreal, October, 1968, The Electrochemical Society, New York, 1969.

[73]   R. H. Kingston, J. Appl. Phys., 27, 101 (1956).

[74]   K. Lehovec and A. Slobodskoy, Solid State Electron., 7, 59 (1969).

[75]   W. Shockley and W. T. Read, Phys. Rev., 87, 835 (1952).

[76]   R. N. Hall, Phys. Rev., 87, 387 (1952).

[77]   D. R. Frankl, Surface Sci., 13, 2 (1969).

[78]   D. R. Frankl, Solid State Electron., 2, 71 (1961).

[79]   L. S. Gasanov and V. I. Stafeev, Soviet Phys.—Semiconductors, 2, 348 (1968).

[80]   H. Statz, G. A. deMars, L. Davis, Jr., and A. Adams, Jr., Phys. Rev., 101, 1272 (1956).

[81]   A. S. Grove, E. H. Snow, B. E. Deal, and C. T. Sah, J. Appl. Phys., 35, 2458 (1964).

[82]   A. S. Grove, B. E. Deal, E. H. Snow, and C. T. Sah, Solid State Electron., 8, 145 (1965).

[83]   A. S. Grove and D. J. Fitzgerald, Solid State Electron., 9, 783 (1966).

[84] K. Lehovec, A. Slobodskoy, and J. L. Sprague, Phys. Status Solidi, 3, 447 (1963).

[85] F. Berz, J. Phys. Chem. Solids, 23, 1795 (1962).

[86] C. G. B. Garrett, Phys. Rev., 107, 478 (1957).

[87] F. Berz, J. Electron. Control, 6, 97 (1959).

[88] A. E. Yunovich, Sov. Phys.—Solid State, 1, 829 (1960).

[89] D. R. Frankl, Surface Sci., 3, 101 (1965).

[90] S. R. Hofstein and G. Warfield, Solid State Electron., 8, 321 (1965).

[91] W. L. Brown, Phys. Rev., 91, 518 (1953).

[92] L. M. Terman, Solid State Electron., 5, 285 (1962).

[93] P. V. Gray and D. M. Brown, Appl. Phys. Lett., 8, 31 (1966).

[94] K. Zaininger and G. Warfield, IEEE Trans. Electron Devices, ED-12, 108 (1965).

[95] E. H. Nicollian and A. Goetzberger, Appl. Phys. Lett., 7, 216 (1965).

[96] F. P. Heiman and G. Warfield, IEEE Trans. Electron Devices, ED-12, 167 (1965).

[97] A. Goetzberger, V. Heine, and E. H. Nicollian, Appl. Phys. Lett., 12, 95 (1968).

[98] J. T. Wallmark and H. Johnson, eds., Field-Effect Transistors, Prentice-Hall, Englewood Cliffs, New Jersey, 1966.

[99] E. H. Snow, A. S. Grove, B. E. Deal, and C. T. Sah, J. Appl. Phys., 36, 1664 (1965).

[100] D. R. Frankl, Surface Sci., 9, 73 (1968).

[101] V. Heine, Surface Sci., 2, 1 (1964).

[102] J. D. Levine and P. Mark, Phys. Rev., 182, 926 (1969). (This paper contains an extensive list of references to earlier work.)

[103] R. F. Greene, D. Bixler, and R. N. Lee, J. Vac. Sci. Technol., 8, 75 (1971).

[104] R. F. Green, Thin Solid Films, 13, 179 (1972).

CHAPTER 3

FIELD EMISSION SPECTROSCOPY OF CHEMISORBED ATOMS

J. W. Gadzuk

E. W. Plummer*

National Bureau of Standards
Washington, D. C.

1. INTRODUCTION

The emission of electrons from a cold metal upon the application of a
strong electric field was one of the earliest confirmations of tunneling as
predicted in the new quantum theory of the 1920's [1-3]. Briefly, field
emission is the process of applying a large electrostatic field, approximately 30,000,000 V/cm, to a cold cathode so that the electrons can tunnel from the metal through the classically forbidden barrier into the vacuum.
In order to achieve these high fields at reasonable voltages, the cathode or
emitter is usually sharpened by a combination of mechanical, chemical,
electrical, or thermal processes to a very sharp point, ~ 1000 Å in radius [4].

*Present address: Department of Physics, University of Pennsylvania,
Philadelphia, Pennsylvania.

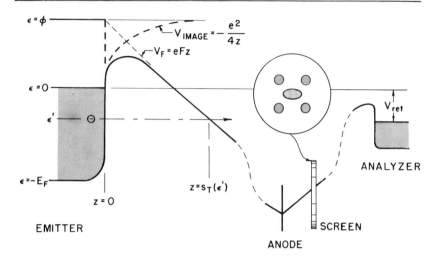

FIG. 1.1. Schematic surface potential and experimental arrangement for a retarding potential energy analysis experiment.

Therefore, several thousand volts applied to an anode will produce the desired field at the emitter surface. The invention of the field emission microscope by Müller [4] made this process a very useful experimental technique. The success Müller obtained with the field emission microscope was a consequence of his realization that if he produced an almost hemispherical tip which was thermally smooth and clean, he could project a greatly enlarged image of the spatial distribution of electrons tunneling from the emitter onto a fluorescent screen. The image on this screen is nearly a stereographic projection of the hemispherical end of the emitter. The dark regions are where the tunneling barrier is higher or wider, i. e., the work function is larger or the field smaller. Likewise bright regions originate because of low work function or high field.

This process is represented by the potential energy diagram in Fig. 1.1. The surface barrier in the presence of the applied field is shown by the dashed curve. Any electron within the metal with energy ε' in an occupied state can then tunnel through the classically forbidden barrier, roughly $0 \leq z \leq s_T$, where s_T depends upon ε'. This process forms the basis of the research area loosely referred to as field emission microscopy [5-12].

The most prevalent type of measurement is a natural consequence of the original theory of Fowler and Nordheim [2, 3]. They derived the relationship, now called the Fowler-Nordheim (FN) equation, relating the current density j from a metal of work function φ with an applied field F:

$$j = AF^2 \exp [-B\varphi^{3/2}/F]$$

with A and B constants. In principle, the slope of $\ln j/F^2$ vs $1/F$ should be a measure of $\varphi^{3/2}$.

It is also possible to measure the distribution in total energy of field-emitted electrons [13-16]. The simplest experimental configuration for such measurements is shown in Fig. 1.1. If a third electrode is placed to the right of the anode (now assumed transparent), then by measuring the collected current as a function of voltage between the collector and emitter and differentiating this quantity with respect to the bias voltage, one hypothetically obtains a retarding potential total energy distribution (TED). This technique has provided a great deal of new information on the electronic properties of clean metal surfaces, some of which has been reviewed by Swanson and Bell [9]. We will therefore not discuss this aspect of field emission.

Field emission has also been a valuable tool for studying partially covered surfaces, chemisorption effects, and film growth. Since 1968, field emission energy distribution (FEED) studies have provided one of the most direct probes of atomic scale adsorption processes at metal surfaces. Both elastic and inelastic processes have been examined. First we briefly mention the classical adsorption studies. As is well known in chemisorption theory, the work function of the composite structure of substrate plus adsorbed partial monolayer seems to be directly related to the coverage on any given crystal face for a given combination of materials [17, 18]. Thus field emission could be used to study chemisorption by measuring the work function from the slope of the FN equation. It is even more reliable for measuring relative work function changes as a function of coverage via FN [19].

Certain problems existed in FEED studies of chemisorption when it was realized that the following picture was breaking down. As envisioned, the

only influence in field emission characteristics of adsorbed atoms should be
through a work function change. However, the slopes of the FN equation and
the TED were often inconsistent with the notion that a work function change
was all that was happening. In a landmark paper, Duke and Alferieff (DA)
pointed out, through idealized model calculations, the importance of resonance tunnelling effects [20]. If an adsorbed atom has a virtual energy level
[21-27] near the Fermi level, then an electron can tunnel through the resonance level of the atom with a much greater transmission probability than
would be accounted for by an effective lowering of the surface barrier. In
fact if the level is sufficiently narrow, and within about 0.5 eV of the Fermi
energy, then the effect should appear as additional structure on a measured
TED. Historically, Sokol'skaya and Mileshkina seem to be the first to have
noticed and to have qualitatively interpreted such structure in their germanium on tungsten experiments [28]. Dobretsov considered their data in terms
of resonance tunneling [29]. In spite of this work and a substantial literature
on resonance tunneling in general [30-32], as well as on resonance tunneling
through impurity states in junctions [33-36], it was not until Duke and Alferieff published their paper that the full impact of resonance tunneling in
field emission was realized. With the advent of resonance tunneling in field
emission, Clark and Young were able to reconcile their strontium-on-tungs-
ten FN and TED data [37]. Plummer, Gadzuk, and Young obtained the first
spectroscopic data relating to virtual levels in adsorbed atoms and presented
[38-43] a method of analyzing resonance tunneling data in a manner similar
to the Fano theory of autoionization [44]. Subsequent discussion has been
due to Modinos and co-workers [45], Gomer [46], and Hurrault [47]. Penn,
Gomer, and Cohen [48] are responsible for a theory of resonance tunneling
similar to that developed in Refs. [38]-[41]. Finally we note that Con-
nors has indicated the similarity between field emission resonance tunneling
and similar phenomena in nuclear, molecular, and chemical physics [49].

Another area of importance with regard to impurity assisted tunneling
has to do with inelastic processes. As demonstrated by Lambe and Jaklevic
in tunnel junctions [50], a tunneling electron can excite vibrational modes of
impurity molecules near one of the interfaces. It is then possible to identify
the energy of these modes by appropriate analysis of the added structure in

the tunneling characteristics [50-53]. Similar studies have been carried out by Swanson and Crouser for the field emission configuration in which adsorbed large organic molecules were considered [54]. Flood has applied the theory of Scalapino and Marcus to inelastic FEED studies [55, 56].

Clark [57] has redone the experiments of Sokol'skaya and Mileshkina [28] in which Ge films are grown on W tips. He has both extended the range of Ge coverages and done this in a much more controlled environment. The experimental results fit in nicely with the new theory due to Nicolaou and Modinos [58].

The hydrogen and oxygen chemisorption experiments of Plummer and Bell are examples of the thorough microscopic picture of atomic events at surfaces which can be obtained through field emission studies [59].

2. GENERAL BACKGROUND

2.1. THEORY

As indicated in the original derivation due to Young [13], the TED of field emitted electrons is obtained in the following manner. The flux of electrons with total energy E and normal energy W incident upon the surface from within the metal is defined as N(E, W). (Note that the free-electron gas approximation is invoked so that the division

$$E = \frac{\hbar}{2m} \left(k_T^2 + k_z^2 \right) = E_T + W$$

can be made. The wave number k_T corresponds to motion in a plane parallel to the surface, whereas k_z corresponds to the normal motion.) The probability per attempt at barrier penetration [$\equiv D(W)$], depends only upon the normal energy. The energy distribution is then

$$\frac{dj}{dE} \equiv j'(E) = \int_0^E N(E, W)D(W)dW \tag{2.1}$$

where the integration is over all normal energies consistent with a total energy E. Roughly the supply or incident flux function is the product of a Fermi function, $f(E) = (\exp(E - E_F)/kT + 1)^{-1}$, and an arrival rate $\frac{1}{\hbar}\frac{\partial E}{\partial k_z}$ the group velocity, times a density of states $\rho(E) \sim 1/\partial E/\partial k_z$ (in one dimension). Harrison [60] has argued that there is an apparent cancellation between the group velocity and the density of states, resulting in a TED of the form

$$j' \sim f(E) \int_0^E D(W)dW \tag{2.2}$$

This is the form originally given by Young [13].

The tunneling probability D(W) has been discussed at great length elsewhere [2, 3, 5, 61], so only end results are cited here. Within the WKB approximation, D(W) is given by an exponential of a phase integral

$$D(W) = \exp\left[-2 \int_{z_1}^{z_2} |k(z)| dz\right] \tag{2.3}$$

with

$$k(z) = \left[\frac{2m}{\hbar^2}(V(z) - W)\right]^{1/2}$$

z_1 and z_2 given by the solutions of $V(z_{1,2}) = W$ and $V(z) = \varphi + E_F - eFz - e^2/4z$, the potential energy associated with the surface barrier. The applied field term is $-eFz$ and the image potential of the surface is $-e^2/4z$. Following standard procedures, Eq. (2.3) can be reduced to the form

$$D(W) \simeq \exp\left(-c + (W - E_F)/d\right) \tag{2.4}$$

with

$$c = \frac{4}{3}\frac{\varphi^{3/2}}{eF}\left(\frac{2m}{\hbar^2}\right)^{1/2} v\left(\frac{\sqrt{e^3 F}}{\varphi}\right) = \frac{0.683\,\varphi^{3/2}}{F} v\left(\frac{3.79F^{1/2}}{\varphi}\right)$$

$$\frac{1}{d} = \frac{2\varphi^{1/2}}{eF}\left(\frac{2m}{\hbar^2}\right)^{1/2} t\left(\frac{\sqrt{e^3 F}}{\varphi}\right) = \frac{1.025\,\varphi^{1/2}}{F} t\left(\frac{3.79F^{1/2}}{\varphi}\right) eV^{-1}$$

where φ is in eV, F in $V/\overset{\circ}{A}$, $W - E_F \ll \varphi$. The functions v and t are given in Table 2.1.

TABLE 2.1

Values of the Functions $v(y)$, $t(y)$

y	$v(y)$	$t(y)$
0	1.0000	1.0000
0.05	0.9948	1.0011
0.1	0.9817	1.0036
0.15	0.9622	1.0070
0.2	0.9370	1.0111
0.25	0.9068	1.0157
0.3	0.8718	1.0207
0.35	0.8323	1.0262
0.4	0.7888	1.0319
0.45	0.7413	1.0378
0.5	0.6900	1.0439
0.55	0.6351	1.0502
0.6	0.5768	1.0565
0.65	0.5152	1.0631
0.7	0.4504	1.0697
0.75	0.3825	1.0765
0.8	0.3117	1.0832
0.85	0.2379	1.0900
0.9	0.1613	1.0969
0.95	0.0820	1.1037
1	0	1.1107

Using the tunneling probability of Eq. (2.4) in Eqs. (2.1) or (2.2) and keeping track of pre-exponential constants yields

$$\frac{dj}{d\epsilon} \equiv j'(\epsilon) = \frac{J_0}{d} f(\epsilon) \exp\left(\frac{\epsilon}{d}\right) \tag{2.5}$$

with

$$J_0 \equiv (4\pi med^2/h^3) \exp(-c) = \frac{1.54 \times 10^{10} \ F^2 \exp(-c)}{\varphi t^2} \ \frac{A}{cm^2}$$

and $\epsilon = E - E_F$. The total current is

$$j_{tot} = \int_{-\infty}^{+\infty} j'(\epsilon)d\epsilon$$

which for zero temperature is simply

$$j_{tot} = J_0$$

the Fowler-Nordheim equation. We will be concerned with deviations from Eq. (2.5) due to the presence of chemisorbed atoms and/or molecules.

Finally we note that the problem of field emission tunneling could be formulated in terms of time-dependent perturbation theory or the transfer Hamiltonian theory [62-64]. In those formulations, the tunneling probability is proportional to the square of a matrix element of some operator T between an initial metal state $|m\rangle$ and final state $|f\rangle$ on the other side of the barrier. Consequently the TED takes the form

$$j'(\epsilon) = \gamma f(\epsilon) |\langle f|T|m\rangle|^2 \tag{2.6}$$

with γ containing constants and slowly varying functions of energy. The utility of Eq. (2.6) is that changes in the TED due to chemisorbed species can more easily be accounted for in the tunneling matrix elements of Eq. (2.6) than in the WKB phase integrals of Eq. (2.3). This fact will enable us to develop a fairly transparent theory of resonance tunneling.

2.2. EXPERIMENT

The heart of the experimental apparatus used in the studies to be discussed here is the energy analyzer. The design begins with the field emission projection geometry [4], where the spatial image of the electrons tunneling from the tip is projected onto a fluorescent screen. A small hole usually called the "probe hole" is cut in the screen. The electrons of interest pass through the probe hole and are then subsequently energy-analyzed. The probe hole is usually stationary and the field emission pattern is deflected until the appropriate crystallographic region is over the probe hole. Since the probe hole samples emission from only a small portion of a facet on the field emitter (~25-100 atoms), the total currents measured are quite small (~10^{-10} A). However, it is then possible to monitor changes in emission characteristics due to adsorption of single atoms.

The first spectroscopic studies of adatoms were done with a retarding potential analyzer designed by Young [39]. Although this analyzer performed nobly, it suffered the same inherent limitation all retarding analyzers have. They measure the integral of the desired energy distribution and the signal must then be differentiated. This is an especially severe limitation in field emission since the number of electrons with low energy is exponentially less than those with higher energies near the Fermi level. Thus a working range of 2 to 3 orders of magnitude in current limits studies to within ~1 eV of the Fermi level.

The adaptation of a deflection energy analyzer to a field emission source has eliminated the shortcomings of the retarding analyzer and increased the current range to 6 to 8 orders of magnitude. The characteristics of such an analyzer have been detailed by Kuyatt and Plummer [65].

The measured TED with adatoms present is most lucidly analyzed by dividing by the clean surface TED to get rid of the dominating but uninteresting exponential. The resultant

$$R(\epsilon) \equiv j'(\epsilon)/j_0'(\epsilon) \tag{2.7}$$

has been termed the enhancement factor. It displays the structure in the TED induced by the adatom energy levels. Since at the present time there is no way of measuring exactly the areas being sampled or the collection efficiency of the analyzer, there is an arbitrary normalization constant introduced in Eq. (2.7). To avoid the effects of this constant, present-day studies plot log R vs ε and thus attach significance only to the position of structure and not to its absolute strength.

3. ADSORPTION STUDIES

The use of TED studies to provide detailed information on a microscopic level about adsorption phenomenon has been particularly successful and interesting. Section 3.1 concerns itself with classical studies in which adsorption is believed to affect a TED only through work function changes. Elastic field emission resonance tunneling is considered in Section 3.2; and inelastic tunneling due to surface impurities is the subject of Section 3.3.

3.1. WORK FUNCTION CHANGES

Unlike FN studies, energy distribution measurements have not been used very much to measure work function changes as a function of coverage of adsorbed atoms. There is a fundamental reason for this. The chemisorbed layer changes not only the barrier height but also its shape and width. This complicated situation can hardly be characterized by a single parameter, the change in work function.

Swanson and Crouser [66] are responsible for the one study in which work functions ascertained from the slope of the energy distribution were measured as a function of adlayer coverage. They studied Ba on (111) Mo and Cs on (100) and (110) W, with a van Oostrom [7] retarding potential analyzer. In general, they observed energy distributions which could be fit to the standard expression, Eq. (2.5), by adjusting the value of φ. By taking TED's at many different coverages (θ), Swanson and Crouser [66] obtained effective φ vs coverage curves which have the usual shape, linearly decreasing φ at low θ dipping to a minimum value of φ less than the work function of a solid of the adsorbate, and then finally rising to the value of the solid adsorbate [18].

However there were many problems with regard to anomalous effective emitting areas, pre-exponentials, and polarizabilities. It was thus clear that work function changes were not the whole story in spite of the superficial agreement with expected φ vs θ behavior, which leads us to the resonance tunneling interpretations.

3.2. ELASTIC RESONANCE TUNNELING

The phenomenon of chemisorption has been approached from many different points of view. The last section utilized what might be called the "macroscopic eye" in that chemisorption was characterized by its effects on material properties such as work functions as opposed to atomic level properties such as energy levels. In recent years theoretical studies of the chemisorbed state have concerned themselves with predicting or interpreting macroscopic quantities in terms of atomic level properties [21-27]. In order that the basis of these theories can be put on a firm foundation, it is necessary to acquire independent information about the atomic states of chemisorbed objects (ions, atoms, molecules). We feel that field emission resonance tunneling has provided a significant tool for such studies, as will be shown. The motivation for the resonance tunneling interpretation resulted largely from experimental problems. As seen in Section 3.1 for many studies of chemisorbed gases on field emitters, the electron emission effects could be simply characterized by a work function change presumably arising from an altered dipole layer at the surface. However there were sufficient inconsistencies in experimental data, such as that of Delchar and Ehrlich [67], to inspire deeper thought on a microscopic level. In particular, Delchar and Ehrlich observed that when nitrogen was adsorbed on the (100) plane of tungsten, not only was the slope of the FN equation reduced (implying a decrease in φ), but the total current was also reduced (implying an increase in φ). More dramatic was the study by Ermrich and van Oostrom in which molecular nitrogen with an excited state roughly 1 eV below the Fermi level was adsorbed and electron bombarded while on or near the (100) plane [68, 69]. In this experiment large FN determined work function increases up to ~4.5 eV were observed with a simultaneous increase in current of up to four orders of magnitude. Clearly some new ingredient to field emission theory was required to understand these unexpected results.

Duke and Alferieff (DA) [20] opened the door to a new era of field emission work by pointing out the role of elastic resonance tunneling through virtual energy levels of atoms or molecules adsorbed on metal surfaces [21, 29]. Essentially they realized that adsorption of an atom cannot only lower or raise the surface barrier resulting in a real work function change but also can change the shape or effective thickness of the barrier due to the presence of the attractive potential well of the adatom. Thus although the thermodynamically defined work function could remain constant, the field emission tunneling probability could drastically change, so that the experimental field emission characteristics, either FN or TED curves, would be nonsensical when the work function is used as the only fitting parameter. In Fig. 3.1 a schematic potential diagram is shown in which the adatom potential is taken to be a square well. If the adatom has a virtual energy level near the Fermi energy, then an electron tunneling from a metal state Ψ_m to a vacuum state Ψ_f can go either via the direct channel or through the intermediate resonance atomic state Ψ_a. The latter case is particularly advantageous for two reasons, both of which tend to greatly enhance the tunneling current. First, the atomic potential cuts a hole in the potential barrier which should reduce the WKB phase integral of Eq. (2.3). Second, by breaking the tunneling process up into two coherent steps, the net tunneling probability is made roughly the sum of the probabilities for each step rather than the product, and this tends to greatly increase the resulting probability, as will be shown [70-72].

In their landmark paper, DA drew upon the substantial body of knowledge dealing with the problem of double barrier penetration [30-34]. Briefly stated, if the energy of the tunneling electron is the same as the energy of a standing wave between the two potential hills, (in our case the valley is the attractive atomic potential) then barrier penetration is much easier. One would hope that the energy of the standing waves in the case of adsorbed atoms would be related to the allowed energy states of the atom, which might be broadened and shifted due to interaction with the solid [21-27]. Thus an energy distribution of electrons field-emitted through the

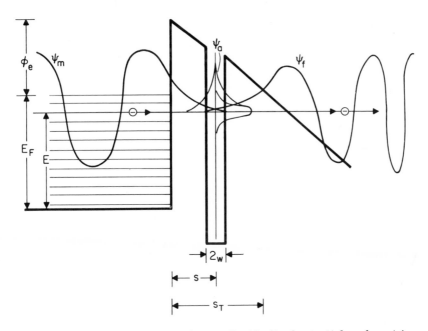

FIG. 3.1. Schematic model showing the idealized potentials relevant in resonance tunneling. The electron wave functions are: Ψ_m, the unperturbed metal function; Ψ_a the localized virtual impurity function; and Ψ_f the emitted electron function.

adatom would tend to display structure which could be related to the energy spectrum or local density of states of the adatom. DA illustrated this point by performing an exact wave–matching calculation for the surface barrier neglecting the image potential, with various combinations of square–well and delta–function model potentials representing the atom core. With their model calculations they also were able to show that neutral adsorbates with repulsive pseudo–potentials could lead to reductions in total current with no change in work function. Because of the numerical nature of the DA theory it was hard to interpret actual experimental TED data in terms of the atomic parameters such as level positions and widths which are the basic ingredients of chemisorption theories [21-27].

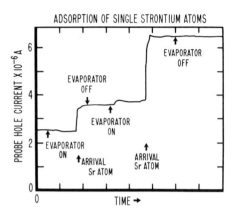

FIG. 3.2. Probe-hole current versus time as the Strontium source is switched on and off. The step increase in current occurs when a single Sr atom arrives upon the surface being viewed.

From the DA theory came the realization that an atomic spectroscopy of adsorbed atoms would be possible. In this pursuit, Clark and Young, following the procedure of Holscher [73], showed in their study of strontium on tungsten [37] that single adatom events could be observed in a field emission microscope. As outlined in Section 2.2, emission from ~30 surface atoms on the desired single crystal face was directed through the probe hole in the fluorescent screen of the anode. Next the Sr source was turned on. When the first Sr atom arrived at the plane viewed by the probe hole, the total current took a step function jump by about a factor of five, corresponding to an enhancement of ~100 in the tunneling current through the Sr atom, since the effective emitting area covered is only ~3 percent of the total area, as shown in Fig. 3.2. At this point the source was turned off and a TED was then measured. The key results of the Clark and Young study relevant to resonance tunneling were that the slope of the FN remained the same while the slope of the TED increased, in accord with DA. This result could be understood if the electropositive adsorbate had a broad energy level whose center was above the substrate Fermi level, as expected for the surface perturbed $5s^2$ ground state configuration of Sr.

The first unambiguous observation of adatom energy levels was given by Plummer, Gadzuk, and Young [38] in their study of zirconium on tungsten. An alternate theory of resonance tunneling was also sketched which enabled closer identification of TED results with atomic spectra. The experimental and theoretical results were greatly extended when barium and calcium were studied [39, 40]. The study of germanium atoms and films has also been completed [41, 57], and will be discussed below.

A fairly transparent theoretical picture has been formulated [38-43] from the point of view implied by Eq. (2.6). In retrospect the essence of the theory can be seen to be quite similar to Appelbaum's theory of s-d exchange tunneling in junctions doped with paramagnetic impurities near one of the interfaces [35, 36, 74]. In a manner similar to that used in the extensive work of Modinos and co-workers [45], we consider tunneling as a scattering phenomenon with the scattering t matrix connecting metal and free states

$$T(\underline{r}) = \mathscr{O}(\underline{r}) + \mathscr{O}(\underline{r}) \frac{1}{E - H + i\delta} T(\underline{r}) \tag{3.1}$$

where $\mathscr{O}(\underline{r})$ must still be specified. From Eq. (2.6), the configuration space matrix elements representing $\langle f|T|m\rangle$ are

$$\Lambda_{M-F} \equiv \int d^3r \, \Psi_f^*(\underline{r}) T(\underline{r}) \Psi_m(\underline{r}) \tag{3.2}$$

with $\Psi_m(\underline{r})$ the exponential tail of the metal wave function in the barrier region and $\Psi_f(\underline{r})$ an Airy function [2, 3, 20]. In the events in which an atomic potential V_a is present in the barrier region, the Hamiltonian in Eq. (3.1) is $H = H_0 + V_s + V_a$ and the Green's function for an intermediate state in which the electron is on the adatom is

$$G_{Atom} = \frac{1}{E - H + i\delta} = \frac{1}{E - H_0 - V_s - V_a + i\delta}$$

For intermediate virtual states on the impurity atom it is reasonable to take $(H_0 + V_s + V_a)\Psi_a = (E_\varphi' + i\Delta)\Psi_a$ with Δ the level width of the atomic virtual state. The level center $E_\varphi' = E_\varphi^0 - eFz$ has been shifted downwards due to the first-order Stark shift of the field. The real space representation of the atom Green's function, is

$$G_{aa}(\underline{r}, \underline{r}'; E_\varphi') = \sum_\varphi \frac{\Psi_{a\varphi}(\underline{r}) \Psi_{a\varphi}^*(\underline{r}')}{E - E_\varphi' - i\Delta} \tag{3.3}$$

From Eqs. (3.1) to (3.3), the configuration space tunneling matrix element is

$$\Lambda_{M-F} = \int d^3 r \Psi_f^*(\underline{r}) \mathcal{O}(\underline{r}) \left\{ \Psi_m(\underline{r}) + \int d^3 r' G_{aa}(\underline{r}, \underline{r}'; E_\varphi') T(\underline{r}') \Psi_m(\underline{r}') \right\}$$

which is alternatively

$$\Lambda_{M-F} = \langle f | \mathcal{O} | m \rangle + \sum_\varphi \frac{\langle f | \mathcal{O} | a, \varphi \rangle \langle a, \varphi | T | m \rangle}{E - E_\varphi' - i\Delta} \tag{3.4}$$

Here the sum on φ is a sum over the complete set of atomic states specified by the set of quantum numbers φ. The first term ($\equiv \Lambda_{M-F}^{(0)}$) is just the amplitude for tunneling directly from the metal to free space through a single action of the generalized tunneling operator $\mathcal{O}$. From Eq. (2.6), it is seen that the standard TED is proportional to the magnitude of the square of $\Lambda_{M-F}^{(0)}$ and thus $j_0' = \gamma |\Lambda_{M-F}^{(0)}|^2$. With an atom on the surface, the TED $j' = \gamma |\Lambda_{M-F}|^2$. As discussed in 2.2, it is convenient to consider the ratio of the energy distributions

$$R(\varepsilon) \equiv \frac{j'(\varepsilon)}{j_0'(\varepsilon)} = \frac{|\Lambda_{M-F}|^2}{|\Lambda_{M-F}^{(0)}|^2} \tag{3.5}$$

for current emitted from the area in which the atom sits.

To illustrate the method, assume that a hypothetical single-level atom is adsorbed. Furthermore, for this exposition, take the transfer matrix elements to be pure real quantities. Then Eqs. (3.4) and (3.5) become

$$R(\varepsilon) = 1 + \frac{T(\varepsilon)^2}{(E - E_\varphi')^2 + \Delta^2} + \frac{2T(\varepsilon)(E - E_\varphi')}{(E - E_\varphi')^2 + \Delta^2}$$

with

$$T(\varepsilon) \equiv \frac{\langle f | \mathcal{O} | a \rangle \langle a | T | m \rangle}{\langle f | \mathcal{O} | m \rangle}$$

The first term corresponds to direct tunneling from metal to vacuum and the second term to the resonance tunneling, while the last term is an interference term between the direct and resonance channels. It is also illustrative to consider the quantity $\Delta j'/j_0' = R(\epsilon) - 1$. Since [22-27] the atomic level width is given by $\Delta \simeq \pi \langle \rho_m(\epsilon_F) \rangle |\langle a|T|m \rangle|^2$, with $\langle \rho_m(\epsilon_F) \rangle$ the metal density of states at the Fermi level, and the virtual impurity density of states is $\rho_a(\epsilon) = \dfrac{1}{\pi} \dfrac{\Delta}{(E - E_\varphi')^2 + \Delta^2}$, it follows that

$$\frac{\Delta j'}{j_0'} = \frac{\rho_a(\epsilon)}{\rho_m} \left| \frac{\langle f|\mathcal{O}|a \rangle}{\langle f|\mathcal{O}|m \rangle} \right|^2 \left[1 + \frac{2(E - E_\varphi')}{T(\epsilon)} \right] \tag{3.6}$$

which shows the strong relationship between the current characteristics and the impurity density of states skewed by the energy-dependent matrix elements. This result, first obtained by Gadzuk, Plummer, Young, and Clark [38-41] has more recently been extended in a much more general theory by Penn, Gomer, and Cohen [48] to include on-site correlation effects within the context of the Anderson model and to include the effects of additional tunneling paths not treated here.

The ratio of tunneling matrix elements can be understood in the manner put forth by Parker and Mead [70]. Consider the multistep tunneling process from metal to atom and then from atom to free space. The current flow from metal to atom is given by

$$j_{m-a} = \gamma(1 - f)P_{m-a}$$

and from atom to vacuum by

$$j_{a-v} = \gamma f P_{a-v}$$

where γ contains constants, f is the probability that an electron is in the atomic state, P_{m-a} is the tunneling probability from metal to atom [essentially $\sim \exp(-2ks)$] and P_{a-v} is the tunneling probability from atom to vacuum [$\sim \exp -2k(s_T - s)$]. Conservation of charge requires $j_{m-a} = j_{a-v}$, which results in $f = P_{m-a}/(P_{m-a} + P_{a-v})$ and the current $j_a = j_{a-v} = j_{m-v} = \gamma P_{a-v}P_{m-a}/(P_{m-a} + P_{a-v})$. Without the atom, tunneling is via a one-step process in which $j_0 = \gamma P_{m-v}$ with $P_{m-v} \simeq \exp(-2ks_T)$. The ratio

$$\frac{j_a}{j_0} = \frac{P_{m-a}P_{a-v}}{(P_{m-a} + P_{a-v})P_{m-v}} = \left| \frac{\langle f|\mathcal{O}|a \rangle}{\langle f|\mathcal{O}|m \rangle} \right|^2 \simeq \exp 2ks \tag{3.7}$$

when $s \ll s_T$ as it is in the field emission case. Furthermore

$$2ks \simeq 2\left[\frac{2m}{\hbar^2}(V_0 - E)\right]^{1/2} s$$

for the triangular barrier, and presumably the same image force corrections and expansions given by Eq. (2.4) can be used here. More detail could be put into this sort of calculation by explicitly choosing an operator $\mathcal{O}$, as the current density operator, the external field, or some other perturbation, and then doing a detailed calculation with atomic wave functions. The result would be similar to the d-band tunneling probabilities discussed by Gadzuk [75] and by Politzer and Cutler [76], i.e., the ratio of resonance tunneling matrix elements from $(n - 1)$d states to ns states would be ~ 0.1 due to the centrifugal barrier confinement. The beauty of the present approach is that we have arrived at the result of Eq. (3.7) without specifying the form of $\mathcal{O}$ or the wave functions. The only restriction has been that the impurity assists tunneling rather than backscattering electrons, a situation discussed by Hurrault [47]. Combining Eqs. (3.6) and (3.7), yields

$$\frac{\Delta j'}{j_0'} = R(\epsilon) - 1$$

$$= \frac{\rho_a(\epsilon)}{\rho_m}\exp(2ks)\left(1 + 2(E - E_\varphi')\exp(-ks)\left(\frac{\pi\rho_m}{\Delta}\right)^{1/2}\right) \qquad (3.8)$$

Although the DA model implicitly contains the same information as Eq. (3.8), the work reported in Refs. [38]-[41] was the first explicit calculation leading to the current enhancement being proportional to the impurity density of states as skewed by an exponential. Connor has also shown that similar behavior is expected in chemical physics when the potential energy barrier for a reactive collision contains a well and in nuclear physics when the fission barrier for heavy nuclei possesses two maxima [49]. Formulas similar to Eq. (3.8) can be derived for multilevel atoms, in which case cross or interference terms between the different levels will further the asymmetry of the measured R factor. We turn now to an examination of some of the systems studied experimentally.

3.2.1. Alkaline Earths

The first major experimental verification of field emission resonance tunneling came in the alkaline earth studies by Plummer and Young [39]

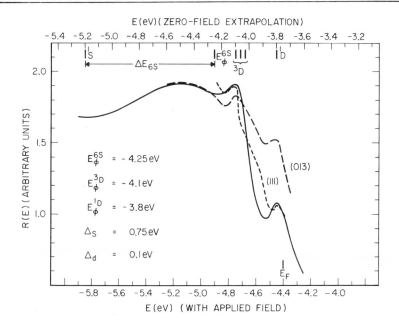

FIG. 3.3. Calculated (solid line) and experimentally obtained (dashed line) values of the enhancement factor, and thus skewed virtual-energy-level spectrum, for Ba on (111) and on (013) W. The magnitude of R is left arbitrary. The scale on the bottom is that of the experiment. The scale on the top is that resulting from an extrapolation to the zero-field limit. Note the positions of the unperturbed atomic levels and the relative shift of the 6s state.

using a retarding potential analyzer. TED measurements with a Ba atom on W revealed several pieces of structure in the enhancement curve which were interpreted in terms of the gas phase excitation spectrum of Ba. Using a formula similar to Eq. (3.8), suitably adapted for the multi-two-electron levels of Ba, a Ba enhancement curve was theoretically determined. Both experimental and theoretical curves are shown in Fig. 3.3. The broad structure at ~-4.5 eV is a manifestation of the $6s^2$ ground state of a Ba atom perturbed and broadened by its interaction with the substrate. The two narrow structures at ~-4.1 and ~-3.8 eV are interpreted to be the first two excited 6s5d states of Ba achieved by promoting a 6s electron into a geometrically more confining 5d orbital. The width of the 6s5d states relative to the $6s^2$

is ~1/8 for the same reasons that the width of the 5d bands in Au are about
10-15 percent of the width of the 6s band, i.e., the confinement by the centrifugal barrier. Similar results were obtained with Ca and Sr adsorption.
Thus a spectroscopy of the adatom density of states, which is characterized
by the level width and position, was truly possible. The general findings
seemed to indicate that ns levels take on a natural width of ~1 eV whereas
the corresponding (n − 1)d levels are ~0.1 eV, down by an order of magnitude.

In this study of Ba, Sr, and Ca adsorption on different faces of tungsten
it appeared that the energy levels of barium on the surfaces shifted from one
face to the next as one would expect solely from a work function change [39].
As the work function increased more of the structure shown in Fig. 3.3 dis-
appeared above the Fermi energy. If the observed levels really didn't depend
on the details of the electron properties of each crystal face of one metal,
would they vary if the work function was constant but the metal changed? Figure 3.4 shows the enhancement curves for single barium atoms on the (110)
plane of molybdenum. The work function of the (110) plane of molybdenum is
~5.00 eV, nearly equal to that of (110) tungsten. It is quite obvious that the
R curves are not the same; in fact we now seem to have two sharp peaks
separated by approximately 0.1 eV, instead of 0.3 eV as in the case of barium on tungsten [39]. The adsorption depends on the substrate, which should
not be a great surprise. What is alarming is that the two peaks which appeared in adsorption on tungsten separated by 0.3 eV are not present. This
could mean that the interpretation of the barium on tungsten curves [39, 40]
is not as general as might be hoped, or that there is structure in the clean
(110) Mo curve which is not present on the tungsten (110) plane. These R
curves were obtained by dividing the after-adsprotion energy distribution
curve by the clean energy distribution [39]. Conceivably this can lead to
errors because of surface sensitive structure in the clean energy distributions. There is also a possibility that the fine structure in both cases arises
from inelastic tunneling, a subject covered in Section 3.3 of this chapter. In
conclusion it seems that a very consistent picture of alkaline earth adsorption
on different faces of W has been constructed but in at least one case of an alkaline earth on a different substrate, specifically Ba on (110) Mo, the description is inadequate.

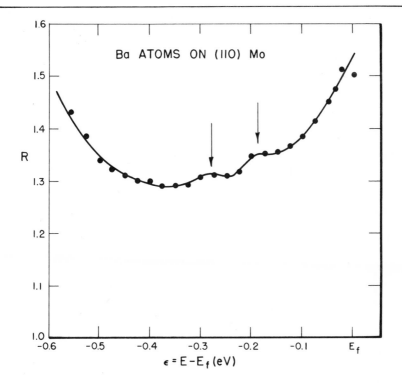

FIG. 3.4. Enhancement factor curve for single barium atoms on (110) Mo. The enhancement factor R in this curve is an average of nearly ten independent curves and is calculated by dividing the energy distribution after adsorption by the energy distribution before adsorption.

3.2.2. Hydrogen on (100) Tungsten

Hydrogen, with only one electron, is undoubtedly the simplest chemisorbed species. For this reason the chemisorption of hydrogen on the cube face of tungsten is an extremely well studied system. The experimental measurements on this system seem to be very reproducible, but unfortunately they do not portray a consistent picture of the hydrogen-tungsten bond. For example, the basic question of whether an observed binding state is atomic or molecular has not been definitely resolved. It is even debatable whether multiple peaks in a flash desorption spectrum indicate multiple binding states upon adsorption or reflect some other mechanism inherent in the flash-off, such as a density-dependent phase transition. This is the type of problem that the spectroscopic information produced from the field emission energy distribution can be used to solve. A detailed study of energy distributions, or specifically the enhancement factor R(E) as a function of hydrogen or deuterium coverage on the (100) plane of tungsten, has been completed [59] utilizing a spherical deflector analyzer [65].

Figure 3.5 displays the enhancement factor R(E) calculated from the measured energy distributions for deuterium and hydrogen on (100) tungsten at 300 K as a function of the density of adsorbed atoms. The characteristic observed low-energy electron diffraction patterns are indicated along the density axis [77]. The free electron energy distributions $j_0{}'(\varepsilon, \varphi, F)$ were calculated using an initial work function of 4.64 eV for the (100) plane and determining the change in work function $(\Delta\varphi)$ from the slope of the Fowler-Nordheim plots [59]. The density of atoms was obtained using Madey and Yates' [78] work function vs coverage data and Tamm and Schmidt's [79] absolute coverage.

The β_2 state saturates at $\sim 5 \times 10^{14}$ atoms/cm^2 and the β_1 state at $\sim 15 \times 10^{14}$ atoms/cm^2 [77-79]. It is clear from Fig. 3.5 that the energy distribution characteristic of β_2 develops with density n from $0 < n < \sim 5 \times 10^{14}$ atoms/cm^2, while the structure characteristic of the clean surface disappears with coverage. The spectrum near $n \simeq 5 \times 10^{14}$ atoms/cm^2 is nearly an ideal example of resonance tunneling [Eq. (3.8)] through a level positioned 0.9 eV below the Fermi energy or 5.8 eV below the vacuum. The full width at half maximum is approximately 0.6 eV. There are additional features in the enhancement curve for the β_2 state. There is a gradual increase in the curve with decreasing energy. This is just the energy-dependent background previously discussed [Eq. (3.8)]. Also there is a shoulder at approximately -1.1 eV below the Fermi energy. This structure could originate from another state, indicating that the energy level in the β_2 is a composite of two levels. In light of the calculations of Newns [25] for hydrogen adsorption and the formalism espoused by Grimley [26], it is

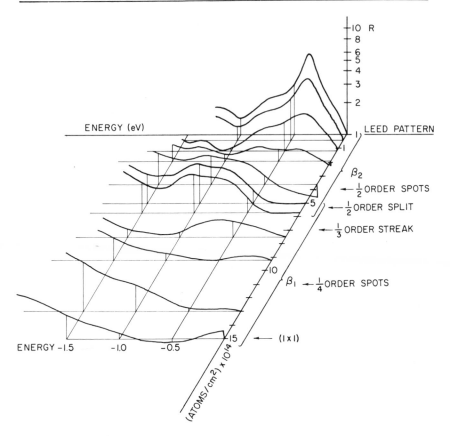

FIG. 3.5. Enhancement factor $R(\varepsilon)$ for hydrogen and deuterium on (100) W at 300 K as a function of the atom density on the surface. The corresponding binding states (β_1 and β_2) and LEED patterns are indicated on the density axis. There are 10^{15} atoms/cm^2 of tungsten in the (100) plane.

tempting to interpret this level as indicating that hydrogen is adsorbed in a nonmagnetic state; that is, the ground state and affinity level come together to form a single–doubly–occupied state. This interpretation must be considered premature at the present time for several reasons: (1) a doubly–occupied level at this position would predict hydrogen existing on the surface as a negative ion; (2) there could be additional structure at lower energies that we cannot see; (3) the level width is narrower than one would expect; and (4) there is hydrogen–hydrogen interaction.

As the coverage increases beyond 5×10^{14} atoms/cm^2 the low-energy electron diffraction (LEED) 1/2 order spots characteristic of the β_2 state begin to split [77], eventually 1/3 order streaks and 1/4 order spots appear before the β_1 is saturated at 15×10^{14} atoms/cm^2 and a simple (1 X 1) structure results. The relative positions of these events have been indicated in Fig. 3.5. The energy distribution characteristic of β_2 disappears rapidly at $\sim 6 \times 10^{14}$ atoms/cm^2, a new level forms which shifts upwards towards the Fermi energy, while a very broad level seems to appear from lower energy as the coverage increases. At saturation two levels are present, a tail of one level extends down from above the Fermi energy and another broad energy level, centered about 7.5 eV below the vacuum level with a width of ~ 2 eV, is seen from below. It is clear from Fig. 3.5 that the β_1 state is not sequentially filling on top of the β_2 state. The energy distribution characteristic of β_2 shifts and converts to the energy level spectrum characteristic of β_1. These distributions indicate that β_2 does not exist on the surface in the presence of β_1 in the same electronic configuration as it has when the coverage is low. These observations would indicate that the multiple peaks in the flash desorption spectrum originate from a density-dependent interaction, specifically a sharp transition near 5 to 6×10^{14} atoms/cm^2. We will discuss the problem of atomic or molecular adsorption in the section on inelastic tunneling.

3.2.3. Oxygen on Tungsten

The chemisorption of oxygen on tungsten is much more complicated than hydrogen adsorption on tungsten. In both systems rapid adsorption occurs up to near monolayer coverage. There hydrogen adsorption seems to quit while in oxygen this rapid region is followed by slow, apparently activated adsorption. Since tungsten oxidizes it is apparent that at some coverage and temperature the surface rearranges or "reconstructs." We would like to show some energy distribution curves from oxygen adsorbed on tungsten to illustrate some of the features discussed above, which do not appear in hydrogen adsorption on tungsten.

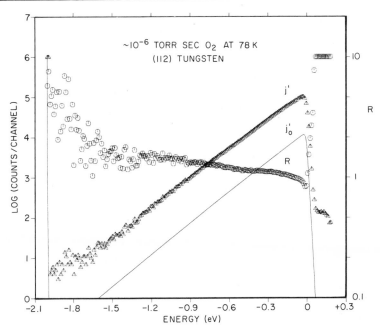

FIG. 3.6.   Example of the data processing system. $j'(\varepsilon)$ (triangles) is
the measured energy distribution for (112) tungsten at 78 K as accumulated
in 256 channels of a multichannel analyzer with 0.010 eV per channel. $j_0(\varepsilon)$
(dotted line) is the calculated energy distribution for a work function $\varphi = 4.90$
eV and a field calculated from the Fowler-Nordheim plot [5]. $R(\varepsilon)$ (circles)
on the right-hand scale is calculated by dividing j' by j_0' for each energy. R
was arbitrarily normalized to equal 1.0 at $\varepsilon = -1$ eV.

The energy distribution of hydrogen on (100) tungsten at a given temperature depends solely upon the density of adatoms, at least for temperatures
above 78 K, and most likely to much lower temperatures.   All that changes
is the sticking coefficient upon adsorption.   In contrast oxygen adsorption at
78 K produces a considerably different energy distribution than adsorption at
room temperature, at least for (110), (112), and (111) planes. So the enhancement factor curves as a function of coverage would look considerably different
at 78 K than at 300 K.   Figures 3.6 and 3.7 show what the enhancement curves

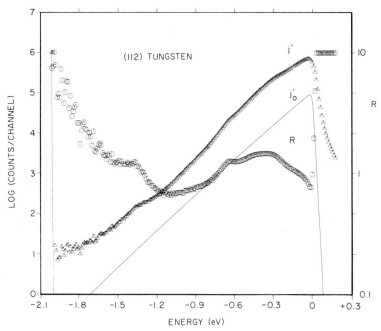

FIG. 3.7. Measured energy distribution j'(ε), calculated free-electron energy distribution $j_0'(ε)$ [φ = 6.14 eV] and enhancement factor R(ε) for 3×10^{-7} torr-sec oxygen exposure on the (112) plane at 78 K.

R(ε) look like for clean (112) tungsten and with 3×10^{-7} torr-sec oxygen exposure at 78 K. Figure 3.8 shows what happens when the substrate is warmed to greater than 200-300 K for 30 sec [80]. In this figure the sample was heated to 1000 K but 300 K was sufficient to produce this irreversible change.

The low-energy electron diffraction data show that oxygen is ordered on the (112) plane at room temperature [81], and may be disordered at 78 K [82]. This irreversible change could be just a disorder-order transition of the adsorbate. The activation energy of diffusion might be too large for ordering to occur at 78 K [83], but the same kind of irreversible transition occurs on the (111) plane slightly below room temperature. Figure 3.9 shows the actual total energy distributions for clean (111) tungsten, saturation exposure of oxygen at 78 K and after warming above 200 K [65]. There is no

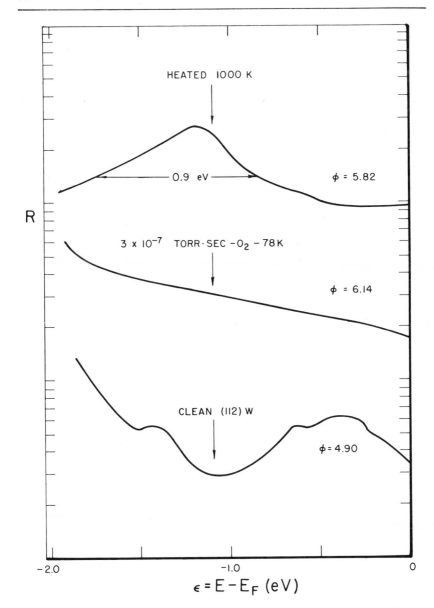

FIG. 3.8. Enhancement factor curves R for clean (112) W; 3 X 10^{-7} torr-sec exposure of oxygen at 78 K; and after the surface had been heated to ~1000 K. Heating to near 300 K produced the same result.

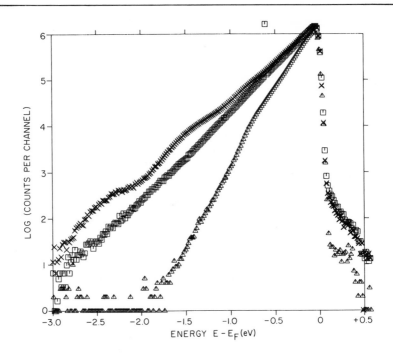

FIG. 3.9. Total energy distribution from (111) W at 78 K. The first energy distribution (triangles) is for clean (111) W; the second (squares) is after saturation exposure of oxygen at 78 K and the third (crosses) is after heating to 250 K for 120 sec.

evidence that the (111) surface orders with oxygen exposure [84]; there does not appear to be any long-range order. This might indicate that the transition we see is from adsorbate to surface molecule, i.e., surface reconstruction, which would not necessarily have to produce ordered low-energy electron diffraction patterns.

We also have included the enhancement curves as a function of exposure from 2×10^{-8} torr-sec to 3×10^{-4} torr-sec of oxygen on the (112) plane of tungsten at room temperature [80] in Fig. 3.10. The point here is that there is considerable action after monolayer coverage ($\sim 5 \times 10^{-6}$ torr-sec). Figure 3.10(a) shows the submonolayer region where one characteristic peak

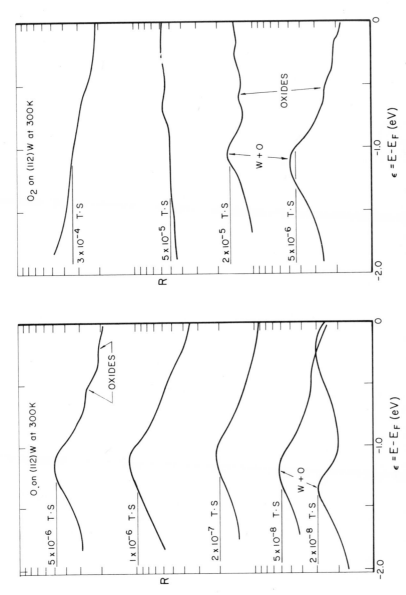

FIG. 3.10.   Enhancement factor curves R(ε) for (112) W at 300 K for various exposures of oxygen.   The multiplicative constant for each curve is arbitrary.

approximately 1.2 eV below the Fermi energy builds in, with coverage up to about 1×10^{-6} torr-sec exposure. Above this two peaks start to build in closer to the Fermi energy. A more detailed study must be made to relate these structures to the LEED patterns and the two peaks observed in the flash desorption spectrum [81]. But the peak at -1.2 eV saturates at about the correct work function to be related to the P(2 X 1) structure and the first adsorption state [81].

Above 1×10^{-6} torr-sec exposure of oxygen [Fig. 3.10(b)] the enhancement curves change gradually until at $> 10^{-4}$ torr-sec there does not appear to be much structure at all. If the substrate is warmed to ~1600 K for 30 sec the enhancement factor curve R reverts back to one resembling a high 10^{-6} torr-sec exposure (see Fig. 3.11). Finally, if the sample is heated to ~2000 K the energy distribution returns to that of the sub-monolayer oxygen coverage (Fig. 3.11). These observations can tentatively be said to agree with the mass spectrometry flash desorption data of King, Madey, and Yates [85] on polycrystalline tungsten. They show that for exposures above 10^{-5} torr-sec tungsten oxides with flash desorption temperatures below 1600 K are adsorbed. While starting at about ~5 X 10^{-6} torr-sec more tightly bound (1700 K) oxides are formed. A much more detailed study of oxygen needs to be undertaken, but these curves serve to illustrate the possibilities.

3.2.4. Krypton on Tungsten

Lea and Gomer [86] performed a series of TED measurements of krypton adsorbed on the (110), (100), (111), and (112) planes of tungsten. Krypton was chosen since in the gas phase it possesses an excited state an energy near the Fermi level of W. Since the perturbation T can mix all the states of the atom, as seen in Eqs. (3.3) and (3.8), the ground state of the adsorbed atom might have structure in its density of states and thus enhancement factors near the excited-state energies. The experiments of Lea and Gomer, in which partial monolayers as well as multilayers of Kr were adsorbed, revealed no pronounced structure in the TED. In most cases it was found that all changes in emission characteristics, both FN and TED curves, could be interpreted in

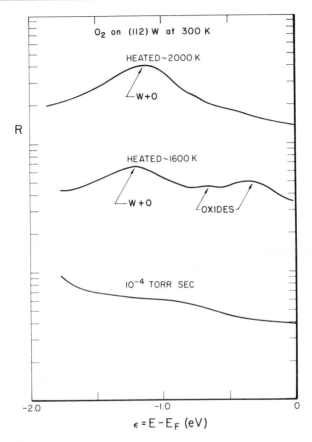

FIG. 3.11. Enhancement factor curve R(ϵ) for oxygen covered (112) tungsten at 300 K.   The lower curve is after 10^{-4} torr-sec exposure of oxygen at 300 K.   The middle curve is after a 30 sec heat to 1600 K.   The top curve is after 30 sec at 2000 K.

terms of field-induced work function decreases. It was found that Kr adsorption on the (100) plane reduced emission from the anomolous structure first observed by Swanson and Crouser [87], further confirming the notion that the structure is due to surface states [88,89]. Finally the role of antiresonances for Kr on (110) W was noted. As pointed out by DA [20] and discussed earlier in this chapter, adsorbates can hinder as well as aid tunneling. In some cases, in which energy levels lie above or far below the Fermi energy, the interference term [the second term in brackets in Eq. (3.8)] can dominate, and for $E < E_\varphi$ the sign is negative, corresponding to the antiresonance. Similar behavior is observed in asymmetric line shapes in atomic resonance phenomenon [44]. Lea and Gomer observed strong decreases in the pre-exponential term in the TED which they feel could not be accounted for by depolarization effects or work function increases. They thus suggest that an antiresonance effect may be the cause of the decrease.

3.2.5. Germanium on Tungsten

Some of the earliest energy distribution work was the Ge on W studies of Mileshkina and Sokol'skaya [28] (MS). The general characteristics they observed were the following. Upon dosing the surface with <1 layer of Ge, the total field emission current was reduced by a factor of ~30-40 (implying an increase in φ), the slope of the FN curve remained constant (implying constant φ), and additional peaks in the TED appeared as the dosage of Ge was increased. With these characteristics, the system Ge on W is a good candidate for elastic resonance tunneling [29].

In a thorough and more recent study of this system Clark [57] has confirmed many of the MS results and greatly extended our knowledge of Ge on W over a wide range of coverage. His results are described quite well by the theoretical model due to Nicolaou and Modinos [58] in which resonance tunneling is the mechanism for changes in emission. A summary of Clark's results are depicted in Fig. 3.12 for various coverages θ in the range $0 \lesssim \theta \lesssim 2$. Energy distributions taken with a retarding analyzer are shown in the top section. As shown, additional structure does not appear until deposition is well into the second layer. Throughout first-layer formation, the pre-exponential A increases, but the work function also increases more strongly and an overall decrease in current is observed. Not until the second layer forms does Ge deposition enhance the emission.

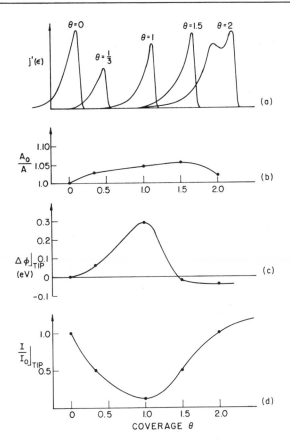

FIG. 3.12. Summary of the Ge on W studies. (a) TED's for various cov-
erages θ; (b) FN pre-exponential vs θ; (c) change of work function vs θ; (d)
normalized probe-hole current vs θ.

These results have been interpreted by Clark in the following way. As the
first layer is deposited, the $4s^2 4p^2$ ^1S ground state of Ge becomes so broad,
due to its interaction with the W substrate, that the level is not distinguishable
in the TED. Furthermore, the Ge layer possesses an overall negative charge,
so the work function increases. For second layer formation, the Ge is further

from the W, so the Ge-W interaction (much stronger than Ge-Ge interactions), is reduced and the line width of the 4p5s derived state is much narrower and appears as structure in the $\theta = 2$ TED in Fig. 3.12. It is believed then that the first layer is characterized by a broad ($\Delta_{FWHM} \simeq 1$ eV) level centered 1.6 eV below the Fermi energy. Not until the second layer is formed does one see evidence for a narrow ($\Delta_{FWHM} \simeq 0.4$ eV) level centered ~0.5 below the Fermi energy. This level is interpreted to be evidence for the 5s first excited state of Ge, somewhat analagous to the 6s5d states seen in the Ba-on-W experiments [39-40]. The broad $4s^2 4p^2$ level is the rudimentary valence band, whereas the 4p5s configuration is probably the first stage of conduction band formation.

3.3. INELASTIC TUNNELING

The importance of impurity assisted inelastic tunneling has been demonstrated in the beautiful junction tunneling experiments of Lambe and Jaklevic [50] (LJ). LJ deposited various molecules and radicals such as H_2O, CO, and C_2H_2 at one interface of a metal-insulator-metal junction. Measurements of the current-voltage characteristics (I-V) revealed sharp line structure in the second derivative curves (d^2I/dv^2). The voltage at which this structure appeared was quite close to the energies of the infrared-active vibrational modes of the molecules or hydroxyl groups. Based upon the results and suggestions of Lambe and Jaklevic, Scalapino and Marcus developed a simple theory to explain these structures in terms of inelastic tunneling [51] in which the tunneling electron excites a vibrational mode of the molecule. With these experiments and theory, a new form of infrared-like spectroscopy was possible.

The corresponding inelastic processes are also possible in a field emission situation. In fact, Flood [55, 56] presented an adaptation of the Scalapino and Marcus theory to the field emission configuration. As will be shown the principal result states that if a localized vibrational mode of energy $\hbar\omega_p$ exists in a molecule adsorbed on the emitter surface, then the resulting FEED could be qualitatively expressed as

$$j_{in}'(\epsilon) = j'(\epsilon) + \sum_{n,p} j'(\epsilon + n\hbar\omega_p)f(\epsilon + n\hbar\omega_p)T_{in}^{(p)} \tag{3.9}$$

with j' the standard TED given by Eq. (2.5); $T_{in}^{(p)} \simeq |\Lambda^{(inel)}(p)|^2$ the inelastic excitation vertex or matrix element; and the sum performed over all modes p and multiples n. This simple result says that replicas of the elastic TED will appear displaced downwards in energy $n\hbar\omega_p$. The relative strength of the inelastic to elastic TED is governed by the electron-phonon coupling $\Lambda^{(inel)}(p)$ which is proportional to the dipole matrix element of the localized vibrator. When this ratio is ~1 percent, as in the case of the LJ experiments and the Plummer-Bell [59] field emission results, then some means for extracting the small inelastic signal, such as second-derivative or computer processing of data is required. On the other hand, in the case of organic molecules studied by Swanson and Crouser [54], the relative strength of inelastic to elastic channels is of order unity or more (which does not exclude 1 percent peaks not reported), so the results are much more dramatic.

The Flood treatment of inelastic field emission starts by assuming a dipole interaction potential between the tunneling electron and the localized vibrator which is

$$V_d(\underline{r}) = -\frac{\mu_z z}{(z^2 + r_T^2)^{3/2}} \tag{3.10}$$

with μ_z the component of the dipole moment of the vibrator normal to the surface. A good deal of effort is then expended to include the interaction term V_d in the WKB phase integral of Eq. (2.3). Calling $\overline{\varphi}_m = \langle \varphi + E_F - (e^2/4z) - eFz - W \rangle$ (the average value in the barrier) and performing the usual tunneling expansions, the phase integral is

$$A(W) = 2 \int_{z_1}^{z_2(\hbar\omega_p)} \left\{ \frac{2m}{\hbar^2}\left[\varphi_m + V_d(\underline{r})\right] \right\}^{1/2} dz$$

$$\simeq 2 \int_{z_1}^{z_2} \left[\frac{2m}{\hbar^2}\varphi_m\right]^{1/2} \left(1 + \frac{V_d(\underline{r})}{2\varphi_m}\right) dz \tag{3.11}$$

Here we note that the classical turning point at z_2, where the electron enters the allowed region in the vacuum, is a function of the energy loss, a point neglected by Flood. For energy losses in the infrared region, this is not too serious, but for losses in the visible region this correction is important and will make the shape of the inelastic portion of the TED different from that of the elastic part. The first term in the integrand of Eq. (3.11) is just the usual WKB result. Thus the exponential part of the WKB tunneling matrix element, can be written

$$\Lambda_{WKB}^{(inel)}(W) = \exp\left(-\frac{c}{2} + \frac{W}{2d}\right)$$

$$\cdot \left\langle \varphi_p \left| \exp\left[-\left(\frac{2m}{\hbar^2}\right)^{1/2} \frac{1}{2\bar{\varphi}_m^{1/2}} \int_{z_1}^{z_2} V_d(\underline{r})dz \right] \right| \varphi_0 \right\rangle$$

where $\bar{\varphi}_m$ is taken to be a constant of order a "few eV's" in the region in which V_d is important and thus is removed from the integral. Since $V_d \ll \bar{\varphi}_m$ in the important region, the exponential in the molecular matrix element is expanded and only the first nonvanishing term is retained. The resulting matrix element becomes

$$\Lambda_{WKB}^{(inel)} = \exp\left(-\frac{c}{2} + \frac{W}{2d}\right)\left(\frac{2m}{\hbar^2}\right)^{1/2} \frac{\xi(\underline{r}_T)}{\bar{\varphi}_m^{1/2}} \left\langle \varphi_p |\mu_z| \varphi_0 \right\rangle \right\}$$

with $\xi(\underline{r}_T) = \left(z_1^2 + r_T^2\right)^{-1/2} - \left(z_2^2 + r_T^2\right)^{1/2}$. The tunneling probability goes as $|\Lambda|^2$, so the resulting TED including both elastic and single-inelastic tunneling is

$$j'(\varepsilon) = \frac{J_0}{d} e^{\varepsilon/d} \left\{ f(\varepsilon) + \sum_p f(\varepsilon + \hbar\omega_p) \left(\frac{2m}{\hbar^2}\right)^{1/2} \frac{2\pi \ln \beta}{\bar{\varphi}_m^{1/2}} |\left\langle \varphi_p |\mu_z| \varphi_0 \right\rangle|^2 \right\} \quad (3.12)$$

with $2\pi \ln \beta$ the value of $\xi(\underline{r}_T)$ averaged over the emitting area, and $\ln \beta \approx \mathcal{O}(1)$. The structure of this final result, which follows from the theory of Scalapino and Marcus [51] and of Flood [56], shows that the TED is the sum of the usual elastic part plus an inelastic replica of the TED at energy $\varepsilon + \hbar\omega_p$. The strength of the inelastic part is directly proportional to the square of the dipole matrix element of the localized oscillator, and it is through this fact that gas-phase results could be of use in the interpretation of inelastic FEED. One point of caution concerns the assumptions leading to Eq. (3.12). The energy dependence of z_2 must be included for large values of $\hbar\omega_p$ [90]. Also modifications to the WKB tunneling probability as discussed by Gadzuk and Plummer [91] must be included when $\hbar\omega_p \simeq |\epsilon| \gtrsim 0.5$ eV. These can lead to nontrivial changes in the shape of the TED.

3.3.1. Organic Molecules

Swanson and Crouser have reported extensive results for the organic molecules phthalocyanine, pentacene, and anthracene deposited on various faces of W and Mo [54]. Their hope was to see and interpret structure in the TED as being due to either resonance tunneling, inelastic vibrational excitation, or inelastic electron excitation within the molecule. The criteria for placing a given peak in one of these categories was taken to be the degree of field dependence of the peak position. Since electron levels at energy E_φ shift with field as $E_\varphi' = E_\varphi + eFz$, $dE_\varphi'/d(eF) = z$ should be of the order of the molecule-solid separation if resonance tunneling is the mechanism. On the other hand, vibrational levels shift very little with field. No a priori rule can be given for electron excitation resonances. A typical set of TED curves [54] for phthalocyanine on the (110) plane of Mo is shown in Fig. 3.13. Plotted in the inset are the shifts of the low-lying peaks as the field is varied. The slopes of these lines are too small to be the result of elastic resonance tunneling, so inelastic tunneling seems to be the mechanism. Although not too much can be said about the shape of the inelastic peaks, they are not inconsistent with a replication of the Fermi-level peak shifted down in energy by the characteristic vibrational frequency. Swanson and Crouser [54] show many more such TED's which are reproducible in the sense that they all have considerable structure but are not reproducible in the sense that the position, relative peak heights, and field shifts bear little resemblance to each other from run to run. As Swanson and Crouser point out this is probably due to a multitude of experimental problems. The current to the probe hole (in their retarding potential analyzer) was occasionally very noisy and erratic due to thermally or field induced random steric changes in the adsorbate-substrate configuration which altered the TED structure. Furthermore it was difficult to position the probe hole over a single molecular spot; thus

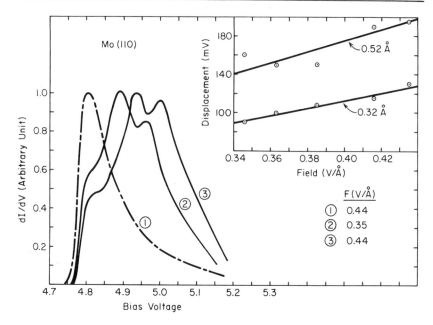

FIG. 3.13. TED spectra for phthalocyanine on Mo (110); curve 1 is clean; inset shows small displacement of peaks with field, indicative of an inelastic loss mechanism (Swanson and Crouser [54]). The bias voltage is given in units of volts.

two or more molecules with different steric configurations may have been contributing to the total TED. Also, the large combination of inelastic and elastic tunneling channels could interfere and couple in manners not understood at this time to produce such results. In any event, these experiments have shown that inelastic FEED is a possible tool for investigating vibrational modes of adsorbed molecules and has pointed the direction for systematic studies of simpler systems.

3.3.2. Hydrogen

The work of Lam and Jaklevic [50] demonstrated that polyatomic molecules in the tunneling barrier are capable of inelastically scattering tunneling electrons in a metal-oxide-metal junction. Swanson and Crouser [54] demonstrated (see Section 3.3.1) that this mechanism of inelastic tunneling is important when polyatomic molecules are adsorbed on a field emitter. An immediate extension of this concept is to simple molecules chemisorbed on specific faces of a field emitter. The problem will now be one of sensitivity, since the cross sections will be small, but the obvious advantages of this type of surface molecular spectrometry, spanning a wide range of wavelengths from the microwave to the visible, is surely worth some effort in signal processing. There were no clear-cut data existing in field emission before the work of Plummer and Bell [59] showing vibrational spectra of simple adsorbed molecules at the metal-vacuum interface. The success in this experiment was due solely to the much improved signal-to-noise ratio of the deflection analyzer [65].

The study of Plummer and Bell [59] centered about determining, from the inelastic tunneling spectrum, whether hydrogen or deuterium was molecularly adsorbed in the β_1 state on the (100) face of tungsten (see Fig. 3.5). If hydrogen was adsorbed as a molecule one would expect to see a step in the R(E) curve at $\epsilon = -\hbar\omega$, where $\hbar\omega$ is the excitation energy of the first vibration mode (~ 0.55 eV for H_2 and ~ 0.4 for D_2). Since the threshold effect in the R(E) curve predicted by Eq. (3.9) will be very small these curves must be amplified by at least a factor of one hundred from those in Fig. 3.5. In other words, the inelastic tunneling effects are much smaller than the tunneling resonance effects in the enhancement curves. Figure 3.14 shows several enhancement factor curves $R(\epsilon)$ for hydrogen and deuterium adsorption on both the (111) and (100) faces of tungsten.

Curve C of Fig. 3.14 is obtained after saturating the (100) plane with deuterium at 300 K and then cooling to 78 K in order to reduce the noise. There is no sign of an inelastic loss at ~ 0.4 eV. If the β_1 state is molecular, then for our signal-to-noise ratio and the coverage given by Tamm and Schmidt [79], the maximum cross section would be $\sim 5 \times 10^{-19}$ cm^{-2}. In order to determine whether the cross section is small or the β_1 state is atomic another state which should be molecular was checked. The γ state on the (111) plane, which builds in below room temperature should definitely be molecular [79, 92]. Curve B of Fig. 3.14 shows the enhancement factor R for deuterium saturation on the (111) plane of tungsten at 78 K. There are at least four

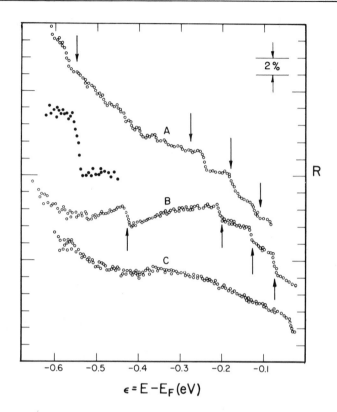

FIG. 3.14. Enhancement factor R(ε) for hydrogen and deuterium adsorption. Curve A is for saturation adsorption of hydrogen on (111) W at 78 K, after the γ state was removed. The segment of a curve indicated by the dark circles is the signal from the molecular γ state of hydrogen. Curve B is for the saturation of deuterium on the (111) plane at 78 K. The molecular vibrational modes are at ~0.4 eV for deuterium and ~0.55 eV for hydrogen on the (100) plane of tungsten at 300 K, indicating no molecular vibrational modes.

losses indicated by the arrows. If the tip is warmed to approximately 200 K the energy loss at 0.4 eV, which is the vibration mode of the molecule, disappears leaving the other modes relatively unchanged. These modes are the deuterium-tungsten modes. The equivalent experiment for the γ state of hydrogen is shown by the segment of the curve in dark circles near 0.55 eV.

Curve A is the hydrogen case when nearly all the γ states are removed. The arrows indicate where the hydrogen-tungsten modes should be if they were shifted by $\sqrt{2}$ from the deuterium case (curve B). The differences between the arrows and the losses in Curve A might result from the fact that the density of hydrogen in Curve A is not the same as the density of deuterium in Curve B.

Excitation of molecularly adsorbed hydrogen and deuterium can be seen and the calculated cross section using Tamm and Schmidt's coverage data is $\sim 4 \times 10^{-18}$ cm^2. If the cross section is the same on the (100) plane, and that is by no means certain, then there is less than 2 percent molecular hydrogen on the (100) plane at saturation coverage at 300 K. The importance of the curves in Fig. 3.14 is that with appropriate signal levels vibrational energy level spectra can be measured for small chemisorbed molecules or atoms. This technique appears promising for the resolution of many questions concerning the chemical nature of species in a given binding state.

ACKNOWLEDGMENT

Portions of this discussion have also appeared in a review article by the authors [93].

REFERENCES

[1] J. R. Oppenheimer, Phys. Rev., 31, 66 (1928).

[2] R. H. Fowler and L. W. Nordheim, Proc. Roy. Soc., Ser. A, 119, 173 (1928).

[3] L. W. Nordheim, Proc. Roy. Soc., Ser. A, 121, 626 (1928).

[4] E. W. Muller, Z. Phys., 106, 541 (1937).

[5] R. H. Good and E. W. Müller, in Handbuch der Physik, Vol. 21 (S. Flügge, ed.), Springer-Verlag, Berlin, 1956, p. 176.

[6] R. Gomer, Field Emission and Field Ionization, Harvard Univ. Press, Cambridge, Mass., 1961.

[7] A. VanOostrom, Philips Res. Rept. Suppl., 11, 102 (1966).

[8] C. W. Todd, Ph.D. Thesis, Cornell Univ., Dept. Appl. Phys. (1971).

[9] L. W. Swanson and A. E. Bell, in Advan. Electron, Electron Phys., 32 (1973).

[10] L. D. Schmidt and R. Gomer, J. Chem. Phys., 43, 2035 (1965).

[11] G. Ehrlich, Ann. Rev. Phys. Chem., 17, 295 (1966).

[12] R. Klein, Surface Sci., 11, 227 (1968); Surface Sci., 11, 430 (1968).

[13] R. D. Young, Phys. Rev., 113, 110 (1959).

[14] M. I. Elinson, F. F. Dobriakova, and F. F. Krapivin, Radiotekhn. i
Elektron, 6, 1342 (1961), [Engl. Transl: Radio Eng. Electron Phys.
USSR, 6, 1191 (1961)].

[15] J. W. Gadzuk, Surface Sci., 15, 466 (1969).

[16] E. W. Müller and T. T. Tsong, Field Ion Microscopy, Principles and
Applications, Elsevier, Amsterdam, 1969.

[17] J. W. Gadzuk, in The Structure and Chemistry of Solid Surfaces
(G. A. Somorjai, ed.), Wiley-Interscience, New York, 1969.

[18] N. D. Lang, Solid State Comm., 9, 1015 (1971).

[19] Z. Sidorski, I. Pelly, and R. Gomer, J. Chem. Phys., 50, 2382 (1969).

[20] C. B. Duke and M. E. Alferieff, J. Chem. Phys., 46, 923 (1967).

[21] R. W. Gurney, Phys. Rev., 47, 479 (1935).

[22] P. W. Anderson, Phys. Rev., 124, 41 (1961); and in Many-Body Physics
(C. DeWitt and R. Balian, eds.), Gordon and Breach, New York, 1969.

[23] A. J. Bennett and L. M. Falicov, Phys. Rev., 151, 512 (1966).

[24] J. W. Gadzuk, Surface Sci., 6, 133 (1967); Surface Sci., 6, 159 (1967).

[25] D. M. Newns, Phys. Rev., 178, 1123 (1969).

[26] T. B. Grimley, J. Vacuum Sci. Technol., 8, 31 (1971).

[27] J. W. Gadzuk, J. K. Hartman, and T. N. Rhodin, Phys. Rev., B, 4,
241 (1971).

[28] I. L. Sokol'skaya and N. V. Mileshkina, Fiz. Tverd. Tela, 3, 3389
(1961); Fiz. Tverd. Tela, 5, 2501 (1963); Fiz. Tverd. Tela, 6, 1786
(1964); Fiz. Tverd. Tela, 7, 1043 (1964); Fiz. Tverd. Tela, 8, 3163
(1966). [Engl. transl.: Sov. Phys. — Solid State, 3, 2460 (1962);
Sov. Phys. — Solid State, 5, 1826 (1964); Sov. Phys. — Solid State,
6, 1401 (1964); Sov. Phys. — Solid State, 7, 838 (1965).]

[29] L. N. Dobretsov, Fiz. Tverd. Tela, 9, 2769 (1967). [Engl. transl.:
Sov. Phys. — Solid State, 9, 2177 (1968).]

[30] D. Bohm, Quantum Theory, Prentice-Hall, Englewood Cliffs, New
Jersey, 1951, pp. 242-295.

[31] E. Merzbacher, Quantum Mechanics, Wiley, New York, 1961, pp. 121-130.

[32] L. V. Iogansen, Zh. Eksperim. i Teor. Fiz., 45, 207 (1963); Zh.
Eksperim. i Teor. Fiz., 47, 270 (1964). [Engl. transl.: Soviet
Phys. —JETP, 18, 146 (1969); Soviet Phys. — JETP, 20, 180 (1965).]
[33] J. C. Penley, Phys. Rev., 128, 596 (1962).
[34] F. W. Schmidlin, J. Appl. Phys., 37, 2823 (1966).
[35] J. Appelbaum, Phys. Rev. Letters, 17, 91 (1966); Phys. Rev., 154,
633 (1967).
[36] P. W. Anderson, Phys. Rev. Letters, 17, 95 (1966).
[37] H. E. Clark and R. D. Young, Surface Sci., 12, 385 (1968).
[38] E. W. Plummer, J. W. Gadzuk, and R. D. Young, Solid State Comm.,
7, 487 (1969).
[39] E. W. Plummer and R. D. Young, Phys. Rev., B, 1, 2088 (1970).
[40] J. W. Gadzuk, Phys. Rev., B, 1, 2110 (1970).
[41] J. W. Gadzuk, E. W. Plummer, H. E. Clark, and R. D. Young, in
Proc. 3rd Mater. Res. Symp., Electronic Density of States (L. H.
Bennett, ed.), NBS Special Report, No. 323, Dec. 1971.
[42] M. L. Glasser, Phys. Rev., B, 3, 1772 (1971).
[43] J. W. Gadzuk, Phys. Rev., B, 3, 1772 (1971).
[44] U. Fano, Phys. Rev., 124, 1866 (1961).
[45] A. Modinos and N. Nicolaou, Surface Sci., 17, 359 (1969); A. Modinos,
Surface Sci., 20, 55 (1970); Surface Sci., 22, 473 (1970); A. Modinos
and A. Theophilou, J. Phys., C., 4, 338 (1971).
[46] R. Gomer, Surface Sci., 22, 445 (1970).
[47] J. P. Hurrault, J. Phys. (Paris), 32, 421 (1971).
[48] D. Penn, R. Gomer, and M. H. Cohen, Phys. Rev. Letters, 27, 26
(1971); Phys. Rev., B5, 768 (1972).
[49] J. N. L. Connor, Phys. Rev., B, 3, 1050 (1971).
[50] R. C. Jaklevic and J. Lambe, Phys. Rev. Letters, 17, 1139 (1966);
J. Lambe and R. C. Jaklevic, Phys. Rev., 165, 821 (1968).
[51] D. J. Scalapino and S. M. Marcus, Phys. Rev. Letters, 18, 459 (1967).
[52] A. J. Bennett, C. B. Duke, and S. D. Silverstein, Phys. Rev., 176,
969 (1968).
[53] J. A. Appelbaum and W. F. Brinkman, Phys. Rev., B, 2, 907 (1970).
[54] L. W. Swanson and L. C. Crouser, Surface Sci., 23, 1 (1970).

[55] D. J. Flood, Phys. Letters, A, 29, 100 (1969).

[56] D. J. Flood, J. Chem. Phys., 52, 1355 (1970).

[57] H. E. Clark, Ph. D. Thesis, American University, Washington, D. C., August, 1971.

[58] N. Nicolaou and A. Modinos, J. Phys., C, 4, 2859 (1971).

[59] E. W. Plummer and A. E. Bell, J. Vacuum Sci. Technol., 9, 583 (1972).

[60] W. A. Harrison, Phys. Rev., 123, 85 (1961).

[61] R. F. Burgess, H. Kroemer, and J. M. Houston, Phys. Rev., 90, 515 (1953).

[62] J. Bardeen, Phys. Rev. Letters, 6, 502 (1961).

[63] M. H. Cohen, L. M. Falicov, and J. C. Phillips, Phys. Rev. Letters, 8, 316 (1962).

[64] C. B. Duke, Tunneling in Solids, Academic Press, New York, 1969.

[65] C. E. Kuyatt and E. W. Plummer, Rev. Sci. Instr., 43, 108 (1972).

[66] L. W. Swanson and L. C. Crouser, in The Structure and Chemistry of Surfaces (G. A. Somorjai, ed.), Wiley-Interscience, New York, 1969.

[67] T. A. Delchar and G. Ehrlich, J. Chem. Phys., 42, 2686 (1965).

[68] W. Ermrich, Philips Res. Rept., 20, 94 (1965).

[69] W. Ermrich and A. van Oostrom, Solid State Comm., 5, 471 (1967).

[70] G. H. Parker and C. A. Mead, Appl. Phys. Letters, 14, 21 (1969).

[71] H. R. Zeller and I. Giaever, Phys. Rev., 181, 789 (1969).

[72] J. W. Gadzuk, J. Appl. Phys., 41, 286 (1970).

[73] A. A. Holscher, Ph. D. Thesis, Univ. of Leiden, 1967.

[74] E. L. Wolf and D. L. Losee, Phys. Rev., B, 2, 3660 (1970).

[75] J. W. Gadzuk, Phys. Rev., 182, 416 (1969).

[76] B. Politzer and P. H. Cutler, Surface Sci., 22, 277 (1970); Mater. Res. Bull., 5, 703 (1970).

[77] P. J. Estrup and J. Anderson, J. Chem. Phys., 45, 2254 (1966); K. Yonehara and L. D. Schmidt, Surface Sci., 25, 238 (1971); D. L. Adams and L. H. Germer, Surface Sci., 23, 419 (1970).

[78] T. E. Madey and J. T. Yates, Jr., in Structure et Propriétés des Surfaces des Solides, Centre National de la Recherche Scientifique, Paris, Publication No. 187, 1970, p. 155.

[79] P. W. Tamm and L. D. Schmidt, J. Chem. Phys., 51, 5352 (1969);
      J. Chem. Phys., 52, 1150 (1970); and further work to be published.

[80] E. W. Plummer, unpublished data given at Physical Electronics Con-
      ference, 1971.

[81] J. C. Tracy and J. M. Blakely, Surface Sci., 15, 257 (1969); D. L.
      Adams, L. H. Germer, and J. W. May, Surface Sci., 22, 45 (1970).

[82] J. T. McKinney, private communication.

[83] T. Engel and R. Gomer, J. Chem. Phys., 52, 1832 (1970).

[84] J. Taylor, Surface Sci., 2, 544 (1964).

[85] D. A. King, T. E. Madey, and J. T. Yates, J. Chem. Phys., 55,
      3236 (1971).

[86] C. Lea and R. Gomer, J. Chem. Phys., 54, 3349 (1971).

[87] L. W. Swanson and L. C. Crouser, Phys. Rev. Letters, 16, 389 (1966);
      Phys. Rev. Letters, 19, 1179 (1967); Phys. Rev., 163, 622 (1967).

[88] E. W. Plummer and J. W. Gadzuk, Phys. Rev. Letters, 25, 1493
      (1970).

[89] J. W. Gadzuk, J. Vacuum Sci. Technol., 9, 591 (1972).

[90] A. D. Brailsford and L. C. Davis, Phys. Rev., B, 2, 1708 (1970).

[91] J. W. Gadzuk and E. W. Plummer, Phys. Rev., B, 3, 2125 (1971).

[92] T. E. Madey, Surface Sci., 29, 571 (1972).

[93] J. W. Gadzuk and E. W. Plummer, "Field Emission Energy Distribu-
      tions," Rev. Mod. Phys., July 1973.

AUTHOR INDEX

Numbers in parentheses are reference numbers and indicate that an author's work is referred to although his name is not cited in the text. Underlined numbers give the page on which the complete reference is listed.

A

Adams, A., Jr., 148(80), 149(80), 162

Adams, D. L., 186(77), 188(77), 190(81), 194(81), 208, 209

Aigrain, P., 108(7), 159

Albers, W. A., Jr., 136, 161

Alferieff, M. E., 138, 140, 162, 168(20), 176, 179(20), 195(20), 206

Allen, R. T., 17, 99

Anderson, J., 186(77), 188(77), 208

Anderson, J. B., 10(21), 98

Anderson, P. W., 168(22, 36), 175(22), 176(22), 177(22), 179(36), 181(22), 206, 207

Anderson, R. L., 142(71), 162

Appelbaum, J. A., 168(35), 169(53), 179(35), 207

Armand, G., 27(76), 35, 91, 100

B

Bardeen, J., 108(10), 112(21), 159, 160, 172(62), 208

Baule, B., 17, 99

Beder, E. C., 5, 11, 14(6), 20, 22, 97, 99

Bell, A. E., 169(59), 186(59), 199, 203, 208

BenDaniel, D. J., 134, 139, 161

Bennett, A. J., 168(23), 169(52), 175(23), 176(23), 177(23), 181(23), 206, 207

Berz, F., 127, 132, 148, 161, 163

Bezak, V., 137(54), 161

Bird, R. B., 5, 97

Bishara, M. N., 26(59), 100

Bixler, D., 157(103), 163

Blakely, J. M., 190(81), 194(81), 209

Blakemore, J. S., 113, 114, 117, 120, 160

Bogan, A., 71(100), 76(110), 77(110), 78(110), 101, 102

Bohm, D., 168(30), 176(30), 206

Boyle, W. S., 120, 160

Brailsford, A. D., 201(90), 209

Brattain, W. H., 108(10), 127, 159, 160

Brinkman, W. F., 169(53), 207

Brown, D. M., 153, 163

Brown, W. L., 152(91), 163

Burgess, R. F., 170(61), 208

Burhop, E. H. S., 20, 99

Busby, M. R., 74, 102

C

Cabrera, N., 22, 23(55), 24(55), 26(55), 27(61), 78(55), 89(61), 99, 100

Calia, V. S., 10(18), 98

Callaway, J., 138(65), 162

Celli, V., 19, 22, 23(49, 55), 24(55), 26(55), 78(55), 99

Cercignani, C., 84(127), 86(127), 103

Chambers, C. M., 27(74), 30, 32, 100

SUBJECT INDEX

A

Acceptors, simple, 116
Accommodation coefficient,
  see under Energy
Accumulation layer, 128
Adsorption
  heat of, 89
  nonequilibrium, 88
  selective, and scattering
    phenomena, 19, 20, 25

B

Barrier shape, see Space charge
Beam source
  aerodynamic nozzle, 7
  Knudsen oven, 8
Bloch functions, 111
Born-Oppenheimer approxima-
  tion, 4

C

Capacitance, surface, 144-145
Capacitively applied electric field,
  definition of, 106
Collision time, 36, 71
Collisions, multiple, 72
Contact,
  semiconductor-insulator, 126
  semiconductor-semiconductor,
    125
Cube models, see Scattering

D

Debye length, see Space charge
Debye temperature, 57
Defect levels, 117

Degeneracies in Fermi-Dirac
  statistics, 114-115
Degenerate bands, 117
Density-of-states function, 118
Depletion layer, 128
Detailed balance, 85
Diffraction of gas atoms, 19
Donors, simple, 116
Drag coefficient, 82

E

Energy accommodation coefficient,
  6, 10, 40
  calculation of, using soft-cube
    model, 59 ff
  experimental data for, 65
  experimental measurements of, 41
  low pressure (LP) method, 10
  modified definition of, 15
  temperature jump (TJ) method, 10
  theoretical predictions of, 66
Energy analyzer, 166, 172 ff
Extrinsic materials, 120

F

Fermi-Dirac distribution function,
  113
Fermi-Dirac integrals, 114
Fermi-Dirac occupation statistics,
  114
Fermi energy, 118
  intrinsic, 119
Field emission, 165
Field emission experimental data
  on Ba/Mo (110), 185
  on Ba/W (111), 183
  on Ba/W (013), 183
  on Ge/W, 196ff

Other books of interest to you...

Because of your interest in our books, we have included the following catalog of books for your convenience.

Any of these books are available on an approval basis. This section has been reprinted in full from our *physical chemistry/physics/surface and colloid chemistry* catalog.

If you wish to receive a complete catalog of MDI books, journals and encyclopedias, please write to us and we will be happy to send you one.

MARCEL DEKKER, INC.
95 Madison Avenue, New York, N.Y. 10016

physical chemistry
physics/surface and
colloid chemistry

ANDERSON *The Raman Effect*

In 2 Volumes

edited by ANTHONY ANDERSON, *University of Waterloo, Ontario*

Vol. 1 *Principles*
416 pages, illustrated. 1971

Vol. 2 *Applications*
640 pages, illustrated. 1973

". . . provides a valuable up-to-date treatment, which is very usefully complementary to previous books on this subject."—N. Sheppard, *Nature*

Provides basic coverage of the theoretical and experimental principles of Raman spectroscopy and discusses its important applications in physics and chemistry. Of value to physical chemists, physicists, and graduate students and other researchers in Raman scattering and related spectroscopic fields.

CONTENTS:

Volume 1: Historical introduction, *R. S. Krishnan.* Polarizability theory of the Raman effect, *G. W. Chantry.* The theory of Raman scattering from crystals, *R. A. Cowley.* Raman instrumentation and techniques, *C. E. Hathaway.* The stimulated Raman effect, *P. Lallemand.* Brillouin scattering, *R. S. Krishnan.*

Volume 2: Applications to inorganic chemistry, *R. Tobias.* Electronic Raman transitions, *J. Konigstein* and *O. Mortensen.* High resolution Raman studies of gases, *A. Weber.* Raman spectra of molecular crystals, *R. Savoie.* Raman spectra of ionic, covalent, and metallic crystals, *G. Wilkinson.*

BLAKER *Geometric Optics:*
The Matrix Theory

by J. WARREN BLAKER, *Department of Physics, Vassar College, Poughkeepsie, New York*

142 pages, illustrated. 1971

Presents a general approach to the study of geometric optics in the paraxial-ray approximation. Offers insight into the physical properties of optical systems, in addition to providing experience in matrix transformations. Designed for students of optics (no previous knowledge of the subject is

needed), and. can be used as a textbook either in or out of the classroom.

CONTENTS: Introduction • Refraction and translation matrices • The optical system; imaging • Additional properties of the system matrix • Mirrors • Stops and pupils; chromatic aberration • Geometric aberrations • Optical instruments.

CHRISTENSEN and IZATT *Handbook of Metal-Ligand Heats and Related Thermodynamic Quantities*

by JAMES J. CHRISTENSEN and REED M. IZATT, *Brigham Young University, Provo, Utah*

336 pages, illustrated. 1970

Provides an extensive compilation of heats of metal-ligand interactions in solution. Especially directed to researchers in coordination chemistry. Also a convenient reference source for workers in the thermochemistry and thermodynamics of metal-ligands interaction, as well as those in such fields as chemistry, physics, bacteriology, engineering, and medicine.

CONTENTS: Use of table and indexes • Table • Author index • Empirical formula index • Element index • Synonym index • Reference index.

COHEN *Statistical Mechanics at the Turn of the Decade*

edited by EZECHIEL G. D. COHEN, *The Rockefeller University, New York City*

246 pages, illustrated. 1971

Surveys the fundamental problems in statistical physics.

CONTENTS: Statistical mechanics and ergodic theory — an expository lecture, *A. Wightman.* The generalization of the Boltzmann equation to higher densities, *E. Cohen.* The C^*-algebra approach to statistical mechanics, *D. Ruelle.* The Curie point, *C. Domb.* Phase transitions in ferromagnets, *F. Dyson.* From the mean field approximation to the method of random fields, *A. Siegert.* A review of superfluids and superconductors, *P. Martin.* Dynamic phenomena near a critical point, *P. Hohenberg.*

CUTLER and DAVIS   Detergency: Theory and Test Methods

In 2 Parts

(Surfactant Science Series, Volume 5)

edited by W. GALE CUTLER, and RICHARD C. DAVIS, *Whirlpool Corporation, Benton Harbor, Michigan*

Part 1 464 pages, illustrated. 1972
Part 2 in preparation. 1973

CONTENTS:

Part I: Introduction, *W. G. Cutler.* Definition of terms, *O. W. Neiditch.* Laundry soils, *W. C. Powe.* Theories of particulate soil adherence and removal, *W. G. Cutler, R. C. Davis, and H. Lange.* Removal of organic soil from fibrous substrates, *H. Schott.* Removal of particulate soil, *H. Schott.* Role of mechanical action in the soil removal process, *B. A. Short.* Soil redeposition, *R. C. Davis.* Evaluation methods for soil removal and soil redeposition, *J. J. Cramer.* Test equipment, *W. G. Spangler.*
Part II: Sequestration, *W. W. Morgenthaler.* Optical brighteners, *P. S. Stensby.* Evaluation of the rinsing process, *W. L. Marple.* Bleaching and stain removal, *C. P. McClain.* Damage to physical substrate, *J. A. Dayvault.* Ancillary tests and methods, *P. Sosis.* Enzymes, *T. Cayle.* Dishwashing, *W. G. Mizuno.* Test methods in toxicology, *L. J. Vinson.* Test methods in dermatology, *C. Gloxhuber and K. H. Schulz.* Cleaning of metals, *E. B. Saubestre.* Detergents and the environment, *M. W. Tenney.*

DAVIS   Electron Diffraction in Gases

by MICHAEL I. DAVIS, *Department of Chemistry, University of Texas, El Paso*

192 pages, illustrated. 1971

The first book devoted exclusively to electron diffraction in gases. Describes both the theories used in the interpretation of gas-phase electron diffraction data as a means of determining molecular structure, and the nature, scope, and limitations of this technique. Designed for graduate students, as well as chemists and physicists doing research in the area of molecular structure.

CONTENTS: An introduction to diffraction phenomena • Introduction to the theory of molecular scattering • Atomic elastic scattering • Inelastic scattering • Scattering by rigid molecules • Scattering by a vibrating molecule • The experiment • Data reduction • Radical distribution function • Molecular scattering curve analyses.

EL BAZ and CASTEL   Graphical Methods of Spin Algebras in Atomic, Nuclear, and Particle Physics

(Theoretical Physics Series, Volume 2)

by EDGARD EL BAZ, *Claude Bernard University of Lyon, France,* and BORIS CASTEL, *Queen's University, Kingston, Ontario*

444 pages, illustrated. 1972

Presents a graphical method used to study angular momentum algebra, and shows how to graphically solve any problem connected with angular momentum coupling in quantum physics. Starts at the fundamental level of quantum mechanics and progresses gradually, making it accessible to the graduate student as well as to the more experienced scientist. Of great value to physicists, applied mathematicians, and physical chemists, and to graduate students in these fields.

CONTENTS: The foundations of the method • Vector addition coefficients • Generalized vector coupling coefficient • Rules governing diagram transformations • The "3nj" coefficients • How to link different coupling modes • Rotation matrices • Spherical harmonics • Irreducible tensor operators (I.T.O.'s) • The two-body interaction • Extraction of the geometrical part of a Feynman diagram • Nuclear electromagnetic interactions • Coefficients of fractional parentage • Angular distribution and polarization phenomena in nuclear reactions • Study of direct nuclear reactions • The three-body problem • The graphical method for SU_3 coupling factors.

ENSMINGER   Ultrasonics: The Low- and High-Intensity Applications

by DALE ENSMINGER, *Mechanical Dynamics Division, Batelle Memorial Institute, Columbus, Ohio*

592 pages, illustrated. 1973

Provides the fundamentals necessary for the understanding of ultrasonics, and is valuable to students interested in this field, engineers involved in the design of ultrasonic systems, and all other potential users concerned with the practical and theoretical applications of ultrasonics.

CONTENTS: Ultrasonics – a broad field • Elastic wave propagation and associated phenomena • Fundamental equations employed in ultrasonic design and applications • Design of ultrasonic horns for processing applications • Basic design of ultrasonic transducers • Determining properties of materials • Nondestructive testing—basic methods and general considerations • Use of ultrasonics in the nondestructive testing of metals • Use of ultrasonics in the inspection of nonmetals • Imaging, process control, and miscellaneous low-intensity applications • Applications of high-intensity ultrasonics—basic mechanisms and effects • Applications of high-intensity ultrasonics based on mechanical effects • Applications of ultrasonics based on chemical effects • Medical applications of ultrasonic energy.

FLOOD   *The Solid-Gas Interface*
In 2 Volumes

edited by E. ALISON FLOOD, *Division of Pure Chemistry, National Research Council, Ottawa, Ontario*

Vol. 1 536 pages, illustrated. 1967
Vol. 2 680 pages, illustrated. 1967

CONTENTS:

Volume 1: Adsorption and eatalysis, *H. Taylor.* The Gibbs and Polanyi thermodynamic descriptions of adsorption, *E. Flood.* Langmuir and BET theories, *S. Brunauer, L. Copeland, and D. Kantro.* Adsorption thermodynamics and experimental measurement, *E. L. Pace.* Heats of adsorption, *J. M. Holmes.* Heats of immersion and the vapor/solid interface, *A. C. Zettlemoyer and K. S. Narayan.* Surface forces and the solid-gas interface, *A. D. Crowell.* Surface energy and surface tension of crystalline solids, *G. C. Benson and K. S. Yun.* Structure of surfaces and its role in adsorption, *W. J. Dunning.* Adsorbate equation of state, *W. A. Steele.* Adsorption theory from the viewpoint of order-disorder theory, *J. M. Honig.* BET surface areas — methods and interpretations, *D. L. Kantro, S. Brunauer, and L. E. Copeland.* Atomically clean solid surfaces — preparation and evaluation, *H. E. Farnsworth.* Physical adsorption at extremely low pressures, *J. P. Hobson.* Homotattic surface and adsorption potentials, *S. Ross.* Commentary on chapters 1 to 15, *G. D. Halsey, Jr.*

Volume 2: The dielectric constant and solid-gas interface, *R. McIntosh.* Magnetic properties of solid-gas interface, *P. Selwood.* Surface and volume flow in porous media, *R. M. Barrer.* Chromatography and solid-gas interface, *H. W. Habgood.* Sorption by liquid-coated solids and gas chromatography, *S. Ross and E. D. Tolles.* Equilibrium swelling due to sorption, *W. Prins.* Some thermodynamic and kinetic factors in sorption, *B. E. Conway.* Mechanical properties and the solid-gas interface, *P. J. Sereda and R. F. Feldman.* Surface potentials and solid-gas interface, *F. C. Tompkins.* Surface chemistry of elemental semiconductors, *A. C. Zettlemoyer and R. D. Iyengar.* Accommodation coefficients and solid-gas interface, *F. M. Devienne.* Simple kinetic theory, and accommodation, reflection, and adsorption of molecules, *E. A. Flood and J. P. Hobson.* Optical properties of solid-gas interface, *J. N. Hodgson.* Infrared spectroscopy and solid-gas interface, *C. H. Amberg.* Application of the techniques of nuclear and electron paramagnetic resonance to study of solid-gas interfaces, *J. G. Aston.* Application of Mössbauer effect to study of surfaces, *M. J. D. Low.* Small-angle x-ray scattering measurements of surface areas, *E. D. Eanes and A. S. Posner.* Active carbon, *J. R. Dacey.* Pore structures, *B. G. Linsen and A. van den Heuvel.* Adsorption hysteresis, *D. H. Everett.* Commentary on chapters 17 to 36, *G. D. Halsey, Jr.*

FONDA and GHIRARDI   *Symmetry Principles in Quantum Physics*
(Theoretical Physics Series, Volume 1)

by LUCIANO FONDA and GIAN CARLO GHIRARDI, *Institute of Theoretical Physics, University of Trieste, Italy*

536 pages, illustrated. 1970

Treats thoroughly the subject of symmetry in modern quantum physics. Aimed at research physicists and graduate students in physics.

CONTENTS: General considerations on the symmetry problem • Geometrical symmetries in ordinary quantum mechanics • A survey of group theory • Internal symmetries in ordinary quantum mechanics • Poincaré group and relativistic equations • Symmetries in quantum field theory • Unitary symmetries.

GOLDFINGER   *Clean Surfaces: Their Preparation and Characterization for Interfacial Studies*

edited by GEORGE GOLDFINGER, *Chemical Division, U. S. Rubber Company, Naugatuck, Connecticut*

408 pages, illustrated. 1970

The first comprehensive collection of papers on the subject of clean surfaces for interfacial studies.

CONTENTS: Surface morphology of cold drawn polyethylene, *A. Peterlin and K. Sakaoku.* The preparation of monodisperse latexes with well-characterized surfaces, *J. W. Vanderhoff, H. J. Van Den Hul, R. J. M. Tausk, and J. Th. G. Overbeek.* Blood-synthetic polymer interface behavior, *H. G. Clark, L. D. Ikenberry, and R. G. Mason.* Ellipsometry as a tool for the characterization of surfaces, *R. R. Stromberg and F. L. McCrackin.* Characterization and preparation of atomically clean surfaces for investigation in ultrahigh vacuum, *H. E. Farnsworth.* Use of the scanning electron microscope in surface characterization, *O. Johari.* The evaluation and production of homotattic solid substrates of certain alkali halides, *S. Ross and J. Hinchen.* Chemisorption studies on clean silver powders using the vacuum ultramicrobalance, *A. W. Czanderna.* Adsorption properties of heterogeneous surfaces, *M. J. Sparnaay.* Surface energy of supersaturated colloidal states, *M. Rosoff and J. H. Schulman.* Electrochemical techniques for the characterization of clean surfaces in solution, *S. Srinivasan and P. N. Sawyer.* The effect of impurities upon the surface potential and the spreading of liquids on mercury, *A. H. Ellison, G. A. Lyerly, and E. W. Otto.* Techniques and criteria in the purification of aqueous surfaces, *K. J. Mysels and A. T. Florence.* The nature of leached glass surfaces, *M. L. Hair.* Pretreatment of mineral surfaces and its effect on their properties, *P. Somasundaran.* The electronic surface states of finite lattices, *P. Mark.*

(continued)

GOLDFINGER *(continued)*

Impurity concentration at "clean" oxide surfaces, *F. M. Fowkes and T. E. Burgess.* The detection and control of organic contaminants on surfaces, *M. L. White.*

GOOD, STROMBERG, and PATRICK
Techniques of Surface and Colloid
Chemistry and Physics

edited by ROBERT J. GOOD, *State University of New York, Buffalo,* R. R. STROMBERG, *National Bureau of Standards, Washington, D.C.,* and R. L. PATRICK, *Alpha Research & Development, Inc., Elverson, Pennsylvania*

Vol. 1 264 pages, illustrated. 1972

Serves as a practical guide to the most important methods used to characterize surfaces and colloids, and their interactions and reactions. Of special value to colloid and surface chemists, as well as to scientists working in related areas.

CONTENTS:

Volume 1: Film balance and the evaluation of intermolecular energies in monolayers, *N. L. Gershfeld.* Measurement of monolayer permeability, *M. Blank.* Ultrafiltration, *C. J. van Oss.* Biomolecular lipid membranes, *H. T. Tien and R. E. Howard.* Methods used in the visualization of concentration gradients, *C. J. van Oss.*

GREEN *Solid State Surface Science*

a series edited by MINO GREEN, *Zenith Radio Research Corporation (U.K.) Ltd., London*

Vol. 1 432 pages, illustrated. 1969
Vol. 2 264 pages, illustrated. 1973
Vol. 3 248 pages, illustrated. 1973

This series fills the need for critical and provocative essays on the properties of the surfaces of solids. Of great benefit to scientists interested in semiconductor science and techniques, and to chemists and physicists interested in catalysis and surface studies.

CONTENTS:

Volume 1: Chemisorbed hydrogen, *J. Horiuti and T. Toya.* Transport and scattering near crystal surfaces, *R. F. Greene.* Simple complexes on semiconductor surfaces, *M. Green and M. J. Lee.* Work function: Measurements and results, *J. S. Rivière.* Epitaxic films of lead chalcogenides and related compounds, *J. N. Zemel.*

Volume 2: Surface thermodynamics of solids, *R. Linford.* Photon scattering from surfaces, *P. Atkins and A. Wilson.*

Volume 3: Theory of gas–surface scattering and accommodation, *R. Logan.* Surface space-charge layers in solids, *D. Frankl.* Field emission spectroscopy of chemisorbed atoms, *J. Gadzuk and E. Plummer.*

HAIR *Infrared Spectroscopy in Surface Chemistry*

by MICHAEL L. HAIR, *Xerox Corporation, Rochester, New York*

336 pages, illustrated. 1967

CONTENTS: Surface chemistry • Infrared spectroscopy: Theory • Application of transmission spectroscopy to surface studies: Experimental considerations • Silica surfaces • Acid oxide surfaces • Adsorption on metals and metal oxides • Some miscellaneous surfaces • Emission, reflection, and Raman spectra • New applications of infrared to surface studies.

HAIR *The Chemistry of Biosurfaces*

In 2 Volumes

edited by MICHAEL L. HAIR, *Xerox Corporation, Rochester, New York.*

Vol. 1 392 pages, illustrated. 1971
Vol. 2 486 pages, illustrated. 1972

"It contains an excellent review of the present status of the fields of surface and colloid chemistry so that anyone . . . wishing to become familiar with them will study this . . . with profit and enjoyment. Prints and reproductions of figures are flawless too."—Andrew Gemant, *Journal of the Electrochemical Society,* March, 1972.

Of utmost importance for surface chemists, physicists, biologists, and medical researchers.

CONTENTS:

Volume 1: Stabilization of macromolecules by hydrophobic bonding: Role of water structure and of chaotropic ions, *W. B. Dandliker and V. A. de Saussure.* Lipids in water, *M. B. Abramson.* Adsorption of biological analog molecules on nonbiological surfaces: Polymers, *B. J. Fontana.* The adsorption of surfactants at solid-water interfaces, *D. W. Fuerstenau.* Adsorption of proteins and lipids to nonbiological surfaces, *J. L. Brash and D. J. Lyman.* Bilayer lipid membranes: An experimental model for biological membranes, *H. T. Tien.* Cellular narcosis and hydrophobic bonding, *L. S. Hersh.*

Volume 2: Interactions between normal and malignant cells, *L. Weiss.* Some aspects of the growth of mammalian cells on glass surfaces, *C. Rappaport.* Surface effects in hemostasis and thrombosis, *E. W. Salzman.* Hemolysis at prosthetic surfaces, *P. L. Blackshear, Jr.* Insolubilized enzymes on inorganic materials, *H. H. Weetall and R. A. Messing.* Insolubilized antigens and antibodies, *H. H. Weetall.* Specific extraction of enzymes using solid surfaces, *J. Porath and L. Sundberg.* Surface properties of calcium phosphates, *C. L. Kibby and W. K. Hall.* Dental adhesives, *G. M. Brauer and E. F. Huget.*

HANLEY Transport Phenomena in Fluids

edited by HOWARD J. M. HANLEY, *Cryogenic Data Center, National Bureau of Standards, Boulder, Colorado*

528 pages, illustrated. 1969

Introduces fundamental concepts concerned with transport and other irreversible phenomena in fluids. Interdisciplinary in scope and of interest to the physicist, physical chemist, chemical engineer, biologist, or anyone who is interested in the properties of fluids in a nonequilibrium state.

CONTENTS: Introduction, *H. Hanley.* Hydrodynamics, *H. Hanley.* Nonequilibrium thermodynamics, *H. Hanley.* Discontinuous systems, *H. Hanley.* Introduction to statistical theory of irreversible processes, *G. Weiss.* Kinetic theory of dilute gases, *E. Cohen.* Kinetic theory of moderately dense gases, *E. Cohen.* Time-correlation functions, *W. Steele.* Some experimental comments on dilute gas transport expressions. *H. Hanley.* Thermal diffusion, *K. E. Grew.* Experimental verification of the Onsager reciprocal relations, *D. Miller.* Biological aspects of transport, *D. Mikulecky.*

HEINEMANN Catalysis Reviews

(Book Edition)

a series edited by HEINZ HEINEMANN, *The M. W. Kellogg Co., Piscataway, New Jersey*

Vol. 1 342 pages, illustrated. 1968
Vol. 2 368 pages, illustrated. 1969
Vol. 3 294 pages, illustrated. 1970
Vol. 4 496 pages, illustrated. 1971
Vol. 5 362 pages, illustrated. 1972
Vol. 6 354 pages, illustrated. 1972
Vol. 7 328 pages, illustrated. 1973

Stimulates new thought and approaches to this broad science. Reviews concern solid state physics and chemistry; electrochemistry and metallurgy; corrosion, polymerization, and biochemistry; and other fields that add to knowledge of catalytic phenomena by analogy or deduction.

CONTENTS:

Volume 1: Zeolites as catalysts. I, *J. Turkevich.* Reactions catalyzed by pentacyanocobaltate (II), *J. Kwiatek.* Reactions of unsaturated ligands in Pd(II) complexes, *E. W. Stern.* Application of computers in chemical reaction systems, *S. S. Grover.* Electronic surface states of ionic lattices, *P. Mark.* Reflectance spectroscopy as a tool for investigating dispersed solids and their surfaces, *K. Klier.* Importance of the electric properties of supports in the carrier effect, *F. Solymosi.* Static volumetric methods for determination of adsorbed amount of gases on clean solid surfaces, *Z. Knor.*

Volume 2: The band picture in the electronic theories of chemisorption on semiconductors, *F. Garcia-Molinar.* Kinetics analysis by digital parameter estimation, *Y. Bard and L. Lapidus.* The surface properties of zeolites as studied by infrared spectroscopy, *D. Yates.* Application of Mössbauer spectroscopy to the study of adsorption and catalysis, *W. Delgass and M. Boudart.* Corrosion of platinum metals and chemisorption, *J. Llopis.* Inhibition processes in the · gas phase, *Z. Szabó.* Determination of heat of adsorption on clean solid surfaces, *S. Černý and V. Ponec.* Prediction of catalytic action as presented in papers before the Fourth International Congress on Catalysis, *V. Kazansky.* United States-Japan conference on catalysis, *J. Turkevich.*

Volume 3: Electrochemical methods for investigating catalysis by semiconductors, *T. Freund and W. Gomes.* Olefin disproportionation, *G. Bailey.* Heat and mass diffusional intrusions in catalytic reactor behavior, *J. Carberry.* The chemical adsorption of sulfur on metals: Thermodynamics and structure, *J. Bénard.* Fixation of molecular nitrogen, *K. Kuchynka.* Olefin polymerization on supported chromium oxide catalysts, *A. Clark.* Catalytic hydrogenolysis over supported metals, *J. Sinfelt.* Chemical relaxation of surface reactions, *G. Parravano.* Identification of rate models for solid catalyzed gaseous reactions, *R. Mezaki and J. Happel.*

Volume 4: Review of ammonia catalysis, *A. Nielsen.* The mechanism of the catalytic oxidation of some organic molecules, *W. M. H. Sachtler.* Equilibrium oxygen transfer at metal oxide surfaces, *G. Parravano.* Isotopic exchange of oxygen ^{18}O between the gaseous phase and oxide catalysts, *J. Nováková.* The use of molecular beams in the study of catalytic surfaces, *R. P. Merrill.* Heterogenous catalysis by electron donor-acceptor complexes of alkali metals, *K. Tamaru.* X-ray photoelectron spectroscopy: A tool for research in catalysis, *W. N. Delgass; T. R. Hughes, and C. S. Fadley.* Electrocatalysis and fuel cells, *A. J. Appleby.* Hydrodesulfurization, *S. C. Schuman and H. Shalit.*

Volume 5: Crystal and ligand field models of solid catalysts, *D. A. Dowden.* The structure and stability of hydrocarbon intermediates on the surface of catalysts, *C. Kemball.* Mechanism and kinetics of some reactions on silicaalumina catalysts, *J. H. de Boer and W. J. Visseren.* Catalytic naphtha reforming, *F. G. Ciapetta and D. N. Wallace.* Catalytic dehydrogenation of C_4 hydrocarbons over chromia-alumina, *S. Carrà and L. Forni.* Auger spectrometry as a tool for surface chemistry, *A. Pentenero.* π-Complex intermediates in homogeneous and heterogeneous catalytic exchange reactions of hydrocarbons and deriva-

(continued)

HEINEMANN (continued)

tives with metals, *J. L. Garnett.* Some applications of nuclear magnetic resonance to surface chemistry, *J. J. Fripiat.* Considerations in the study of reaction sets, *A. H. Weiss.*

Volume 6: Some aspects of catalysis: The P. H. Emmett Award address, *R. J. Kokes.* A temperature programmed desorption technique for investigation of practical catalysts, *R. J. Cvetanović and Y. Amenomiya.* Recent developments in hydroformylation catalysis, *F. E. Paulik.* On the mechanism of the oxo reaction, *M. Orchin and W. Rupilius.* Electron localization and oxygen transfer reactions on zinc oxide, *Ph. Roussel and S. J. Teichner.* Study of kinetic structure using marked atoms, *J. Happel.* Catalysis for control of automotive emissions, *F. G. Dwyer.* X-ray scattering techniques in the study of amorphous catalysts, *P. Ratnasamy and A. J. Léonard.*

Volume 7: The use of isotopic tracers in studying catalysts and catalytic reactions, *P. Emmett.* The uses of deuterium in the study of heterogeneous catalysis, *R. Burwell, Jr.* Catalytic decomposition of formic acid on metal oxides, *J. Trillo, G. Munuera, and J. Criado.* The surface structure of and catalysis by platinum single crystal surfaces, *G. Somorjai.* NMR studies in adsorption and heterogeneous catalysis, *E. Derouane, J. Fraissard, J. Fripiat, and W. Stone.* The structure and catalytic properties of palladium-silver and palladium-gold alloys, *E. Allison and G. Bond.* Catalysis on supported metals, *A. Cinneide and J. Clarke.* The Fifth International Congress on Catalysis, *R. Burwell, Jr.*

HELFFERICH and KLEIN
Multicomponent Chromatography:
Theory of Interference

(Chromatographic Science Series,
Volume 4)

by FRIEDRICH HELFFERICH, *Shell Development Co., Emeryville, California* and GERHARD KLEIN, *University of California, Richmond*

432 pages, illustrated. 1970

The first comprehensive treatment of chromatographic behavior in systems in which the various sorbable species affect one another. Clarifies the nature and consequences of interference in chromatography, building on a new concept, "coherence," that appears to be broadly applicable to multicomponent dynamic systems in many areas of physics and engineering.

CONTENTS: Introduction • Definitions and basic concepts • Theoretical basis • Column response • Relaxation of simplifying premises • Assessment and outlook.

HOLM and BERRY *Manual on*
Radiation Dosimetry

edited by NIELS W. HOLM, *Danish Atomic Energy Commission, Roskilde, Denmark,* and ROGER J. BERRY, *Churchill Hospital, Oxford, England*

472 pages, illustrated. 1970

Combines fundamental background of the subject with sets of step-by-step procedures. For medical physicists, research workers using radiation techniques, radiobiologists, and operators of industrial radiation plants.

CONTENTS: Introduction to basic concepts and principles in radiation dosimetry, *R. J. Berry and N. W. Holm.* The theory of dosimeter response with particular reference to ionization chambers, *T. E. Burlin.* Calorimetry, *B. Radak and V. Marković.* Aqueous chemical dosimetry, *N. W. Holm and Z. P. Zagorski.* Radiophotoluminescent and thermoluminescent dosimetry, *J. R. Cameron.* Films, dyes, and photographic systems, *W. L. McLaughlin.* Dosimetry in radiation protection, *R. Oliver.* Dosimetry in medical radiation therapy, *L. H. Lanzl.* Cobalt-60 dosimetry in radiation research and processing, *J. Weiss and F. X. Rizzo.* Dosimetry in accelerator research and processing, *E. M. Fielden and N. W. Holm.* The Fricke dosimeter, *K. Sehested.* The ferrous-cupric dosimeter, *E. Bjergbakke.* The ceric sulphate dosimeter, *E. Bjergbakke.* The water dosimeter, *E. J. Hart.* The hydrated electron dosimeter, *E. J. Hart and E. Fielden.* The oxalic acid dosimeter, *N. W. Holm.* The benzene-water dosimeter, *T. R. Johnson.* The ethanol-chlorobenzene dosimeter, *I. Dvornik.* Rigid vinyl film dosimeter, *C. Artandi.* The clear PMMA dosimeter, *C. G. Orton.* Red perspex dosimetry, *B. Whittaker.* Cinemoid color films, *N. Goldstein.* Radiochromic dye-cyanide dosimeters, *W. L. McLaughlin.* Photographic film dosimeters, *W. L. McLaughlin.* Thermoluminescence dosimetry with calcium fluoride, *F. H. Attix.* Lithium fluoride thermoluminescent dosimetry, *J. R. Cameron.* The hydrogen pressure dosimeter, *R. Sheldon.* Radioluminescent dosimetry system, *S. J. Malsky, B. Roswit, C. B. Reid, and C. G. Amato.* The "n" on "p" solar-cell dose-rate meter, *A. C. Muller.* The "p" on "n" solar cell integrating dosimeter, *A. C. Muller.*

JUNGERMANN *Cationic Surfactants*

(Surfactant Science Series, Volume 4)

edited by ·ERIC JUNGERMANN, *Armour-Dial, Inc., Chicago, Illinois*

672 pages, illustrated. 1970

Provides both the industrial and academic research workers with up-to-date coverage and a critical review of four major disciplines covering the field of cationic surfactants: organic chemistry, physical chemistry, analytical chemistry, and biology. Novel concepts are proposed and new experimental procedures are described.

CONTENTS: Introduction, *E. Jungermann.* Straight-chain alkylammonium compounds, *W. M. Linfield.* Cyclical alkylammonium compounds, *A. J. Wysocki and D. Taber.* Petroleum derived cationics, *H. E. Tiefenthal, E. J. Miller, Jr., and P. L. Du Brow.* Polymeric cationic surfactants, *K. Longley.* Miscellaneous non-nitrogen-containing cationic surfactants, *W. M. Linfield, and B. E. Edwards.* Micelle formation of cationic surfactants in aqueous media, *E. W. Anacker.* Micelle formation of cationic surfactants in nonaqueous media, *A. Kitahara.* Absorption of cationic surfactants by cellulosic substrates, *H. J. White, Jr.* Adsorption of cationic surfactants on mineral substrates, *M. E. Ginn.* Adsorption of cationic surfactants on miscellaneous solid substrates, *M. E. Ginn.* Coacervation in cationic surfactant solutions, *A. E. Vassiliades.* A critical review of techniques for the identification and determination of cationic surfactants, *J. T. Cross.* Germicidal properties of cationic surfactants, *C. A. Lawrence.* Toxicology of cationic surfactants, *R. A. Cutler and H. P. Drobeck.*

LAMOLA Creation and Detection of the Excited State

In 2 Parts

(Creation and Detection of the Excited State Series, Volume 1)

edited by ANGELO LAMOLA, *Bell Telephone Laboratories, Murray Hill, New Jersey*

Part A 392 pages, illustrated. 1971
Part B 302 pages, illustrated. 1971

Brings together, for the first time, critical discussions of all the methods of creating, detecting, and characterizing molecular excited states. Of special value to graduate students and working scientists in the fields of molecular spectroscopy, photochemistry, radiation chemistry, photobiology, and radiation biology.

CONTENTS:

Part A: Special methods in absorption spectrophotometry, *W. G. Herkstroeter.* Polarized light in spectroscopy and photochemistry, *F. Dörr.* Quantum yields and kinetics of photochemical reactions in solution, *H. E. Johns.* Energy transfer kinetics in solution, *P. J. Wagner.* Transient luminescence measurements, *W. R. Ware.* Electron spin resonance of excited triplet states, *M. Guéron.* Luminescence spectroscopy, *J. W. Longworth.*

Part B: Electron impact, *M. A. Dillon.* Ionizing radiation, instrumentation, and methods, *R. H. Johnsen.* Observation of excited states by pulse radiolysis, *J. K. Thomas.* Vacuum ultraviolet techniques in photochemistry, *J. R. McNesby, W. Braun, and J. Ball.* Vacuum techniques in

photochemistry, *H. E. Gunning and O. P. Strausz.* Preparative photochemical instrumentation and methods, *W. M. Hardham.*

LEVINE and DEMARIA Lasers: A Series of Advances

edited by ALBERT K. LEVINE, *Division of Science and Engineering, Richmond College, Staten Island, New York,* and ANTHONY J. DEMARIA, *Quantum Physics Laboratory, United Aircraft Research Laboratories, East Hartford, Connecticut*

Vol. 1 384 pages, illustrated. 1966
Vol. 2 456 pages, illustrated. 1968
Vol. 3 384 pages, illustrated. 1971

A series of critical reviews which provides the background, principles, and working information needed by physical and biological scientists seeking to use lasers in research, engineers developing lasers for commercial and military applications, and specialists in any one aspect of lasers who wish to bring themselves up to date authoritatively in the other aspects.

CONTENTS:

Volume 1: Pulsed ruby lasers, *V. Evtuhov and J. K. Neeland.* Optically pumped pulsed crystal lasers other than ruby, *L. F. Johnson.* Organic laser systems, *A. Lempicki and H. Samelson.* Q modulation of lasers, *R. W. Hellwarth.* Modes in optical resonators, *H. Kogelnik.*

Volume 2: Gas lasers, *C. K. N. Patel.* Glass lasers, *E. Snitzer and C. G. Young.* The injection laser, *W. P. Dumke.* Nonlinear optics, *R. W. Terhune and P. D. Maker.* Frequency stabilization of gas lasers, *T. G. Polanyi and I. Tobias.*

Volume 3: Semiconductor lasers, *H. Kressel,* CO_2 lasers, *P. Cheo,* Dye lasers, *M. Bass, T. Deutsch, and M. Weber.*

LINFIELD Anionic Surfactants

(Surfactant Science Series)

edited by WARNER M. LINFIELD, *U.S. Department of Agriculture, Philadelphia, Pennsylvania*

in preparation. 1974

CONTENTS (tentative): Introduction, *W. Linfield.* Petroleum derived raw materials, *G. Hinds.* Lipid and other non-petrochemical raw materials, *F. Scholnick.* Reaction mechanisms of sulfation and sulfonation, *B. Edwards.* Alcohol and ether alcohol sulfates, *S. Shore and D. Berger.* Sulfated monoglycerides and alkanolamides, *J. Weil and A. Stirton.* Sulfated fats and oils, *B. Dombrow.* Alkylaryl sulfonates, *G. Feighner.* Petroleum sulfonates, *C. Bluestein and B. Bluestein.* Alpha-olefin sulfonates, *H. Green.* Derivatives of alpha-sulfomonocarboxy-

---

† *Volume edited by Albert K. Levine*

(continued)

LINFIELD *(continued)*

lic acids, *T. Micich*. Sulfoethyl esters and amides of fatty acids, *L. Burnette*. Glyceryl ether sulfonates, *D. Whyte*. Phosphorus containing anionic surfactants, *E. Jungermann and H. Silbermann*. Acylated amino acids and other anionic surfactants, *J. Spivack*.

LISSANT Emulsions and Emulsion Technology

edited by KENNETH J. LISSANT, *Petrolite Corporation, Saint Louis, Missouri*

in preparation. 1974

Covers the theoretical aspects of emulsions and emulsion preparation from both the physical and geometric points of view. Discusses the types of emulsions used, the reasons for their use, how they are prepared, and how they serve their functions in specific industrial areas.

CONTENTS (tentative): Basic theory, *K. Lissant*. Making and breaking emulsions, *K. Lissant*. Micro emulsions, *L. Prince*. Agricultural emulsions, *P. Lindner*. Food emulsions, *M. Lynch and W. Griffin*. Pharmaceutical emulsions, *B. Mulley*. Emulsion polymerization, *T. Matsumoto*. Emulsions in the paint industry, *G. Allyn*. Asphalt emulsions, *R. Ferm*. Emulsions in the papermaking industry, *F. Vaurio*. Emulsions in printing and the graphic arts, *J. Bulloff*. Hydraulic fluid emulsions, *R. Holzmann*. Cutting oil emulsions, *R. Holzmann*. Emulsions in the cosmetic industry, *C. Fox*.

McINTOSH Dielectric Behavior of Physically Adsorbed Gases

by ROBERT L. MCINTOSH, *Department of Chemistry, Queen's University, Kingston, Ontario*

176 pages, illustrated. 1966

CONTENTS: General introduction • The calculation of polarization of the adsorbate • Variations of ϵ' and ϵ'' with frequency and temperature • Experimental • Experimental results • Effect of sorbed matter on polymers, cellulose, starch, and proteins • Adsorbates on nonporous solids.

MAMANTOV Molten Salts: Characterization and Analysis

edited by GLEB MAMANTOV, *Department of Chemistry, University of Tennessee, Knoxville*

632 pages, illustrated. 1969

A collection of papers of a review-and-research nature on most aspects of molten salt chemistry, and the first book to emphasize characterization and analysis as well as solution chemistry in molten salts. Of interest to both research workers in molten salts and to those in related areas such as metallurgy, glass technology, and solid state science.

CONTENTS: Some fundamental concepts in the chemistry of molten salts, *M. Blander*. The experimental evidence for "complex ions" in some molten salt mixtures, *M. A. Bredig*. The role of phase equilibria in molten salt research, *R. E. Thoma*. Ionic interactions in molten salts, *N. H. Nachtrieb*. Coordination equilibria of nickel(II) in molten chloride salts, *G. P. Smith, J. Brynestad, C. R. Boston, and W. E. Smith*. Electronic absorption spectra and the nature of dilute solutions of alkali metals in fused alkali halides, *D. M. Gruen, M. Krumpelt, and I. Johnson*. Spectrophotometric studies of solute species in molten fluoride media, *J. P. Young*. Advances in the infrared spectroscopy of molten salts, *J. P. Devlin, P. C. Li, and R. P. J. Cooney*. A review of Raman spectroscopy of fused salts and studies of some halide-containing systems, *V. A. Maroni and E. J. Cairns*. The investigation of fused salts by the inelastic scattering of cold neutrons, *J. K. Wilmshurst and J. M. Bracker*. Transport processes in low-melting molten salt systems, *C. A. Angell and C. T. Moynihan*. Transport properties: Conductivity, viscosity, and ultrasonic relaxation, *P. B. Macedo and R. A. Weiler*. Electronic conduction in fused salts, *L. F. Grantham and S. J. Yosim*. Electrical conductivity measurements in molten fluoride mixtures and some general considerations on frequency dispersion, *G. D. Robbins and J. Braunstein*. The distribution of metals between molten fluorides and liquid metals, *D. M. Moulton, W. R. Grimes and J. H. Shaffer*. Investigation of the capacity of the electrical double layer in molten chlorides using a dropping metal electrode, *R. J. Heus, T. Tidwell, and J. J. Egan*. Solutions of halogens in molten halides, *J. D. Van Norman and R. J. Tivers*. Electrode reactions in molten fluorides, *G. Mamantov*. Voltammetric studies of chromium(II) in molten LiF-BeF$_2$-ZrF$_4$ at 500°C, *D. Manning and J. Dale*. Thermochemistry, complexation, electron and oxygen transfer in fused nitrates, *J. Jordan, W. B. McCarthy, and P. G. Zambonin*.

MATAGA and KUBOTA Molecular Interactions and Electronic Spectra

by NOBORU MATAGA, *Department of Chemistry, Osaka University, Osaka, Japan* and TANEKAZU KUBOTA, *Shionogi Research Laboratory, Osaka, Japan*

520 pages, illustrated. 1970

Summarizes all important topics of molecular interaction, such as hydrogen bonding and charge transfer. In addition, there is an elementary description and summary of quantum theories of molecular electronic structures and electronic transition processes. Of interest to students and researchers involved in the fields of molecular electronic spectra and molecular interaction.

CONTENTS: Elements of quantum mechanics • Fundamentals of molecular electronic states • Radiation transition probabilities • Radiationless transitions in molecules • Intermolecular electronic excitation transfer • Charge transfer complexes • Hydrogen bonding complexes • Solvent effect on the electronic spectra • 'Excimers and related phenomena.

MATTSON and MARK
Activated Carbon: Surface Chemistry and Adsorption from Solution

by JAMES S. MATTSON, *Rosenstiel School of Marine and Atmospheric Sciences, University of Miami, Florida,* and HARRY B. MARK, JR., *Department of Chemistry, University of Cincinnati, Ohio*

248 pages, illustrated. 1971

Critically reviews the body of literature available on the surface chemistry of activated carbon, and offers a clear picture of the carbon surface oxides and the role of the carbon surface in solution adsorption mechanisms. Highly recommended to all chemists, technicians, and students in water and wastewater treatment; pharmaceutical and organic chemical purification; sanitary, civil, and chemical engineering and the sugar industry.

CONTENTS: Introduction • Activation of carbon • Surface oxygen functional groups and neutralization of base by acidic surface oxides • Spectroscopic methods for molecular structure determinations on the surfaces of activated carbons • Nature of the electrical double layer • Adsorption of electrolytes • Adsorption of weak and nonelectrolytes from aqueous solution.

MELTON   *Principles of Mass Spectrometry and Negative Ions*

by CHARLES E. MELTON, *Department of Chemistry, University of Georgia, Athens*

328 pages, illustrated. 1970

The first *textbook* presenting not only the physical concepts of mass spectroscopy but also a thorough mathematical treatment. Well organized and developed for teaching both chemistry and physics students at the undergraduate and graduate levels. Also of value to professionals engaged in pure and applied research.

CONTENTS: Introduction • Motion of charged particles in fields • Ion sources • Mass analysis • Ion detectors • Positive ions • Negative ions • Secondary reactions.

MUROV   *Handbook of Photochemistry*

by STEVEN L. MUROV, *Department of Chemistry, State University of New York, Stony Brook*

in preparation. 1973

Contains an extensive compilation of photochemical data, which includes information on ground and excited state properties of organic molecules, properties of useful solvents and low temperature organic glasses, and photochemical techniques and apparatus. Also provides a reference and bibliography section on other literature in the field. Especially directed to photochemists, organic chemists, and molecular spectroscopists.

CONTENTS (tentative): Spectroscopic properties of sensitizers and quenchers • Triplet energies of organic compounds — arranged in order of increasing triplet energy • Lifetimes and quantum yields of phosphorescence at 77°K. • Quntum yields of intersystem crossing in solution • Rates of energy transfer and solvent viscosities • Relative rates of hydrogen abstraction by benzophenone • Three kinetic schemes • Bond dissociation energies • Solvent properties • Transmission characteristics of light filters • Spectra of various glasses • Spectral distribution of ultraviolet sources • Actinometry • Photolysis apparatus • Photomultiplier response • Cooling baths • Dipole moments of excited states • ESR parameters of triplet states • Atomic weights and spin-orbital coupling constants • Fundamental constants and conversion factors • Infrared group frequencies • Nuclear magnetic resonance chemical shift parameters • Boiling points at reduced pressure • Reduction potentials of aromatic hydrocarbons • Ionization potentials • Guide to the photochemical literature and bibliography.

O'KONSKI   *Molecular Electro-Optics*

edited by CHESTER T. O'KONSKI, *Department of 'Chemistry, University of California, Berkeley*

in preparation. 1974

Comprehensively deals with the characterization of molecules by the use of electro-optic relaxation methods. Of interest to physicists, physical chemists, polymer chemists, biophysicists, molecular biologists, biochemists, and electronics engineers.

CONTENTS: Discovery and early explorations of electro-optic effects, *C. .O'Konski.* Electric birefringence of gases and liquids, *A. Buckingham.* Electric birefringence a'nd relaxation in soultions of rigid macromolecules, *C. O'Konski and S. Krause.* Theory of rotational diffusion of suspensions, *D. Ridgeway.* Electric birefringence of flexible polymers, *R. Jernigan and*

(continued)

O'KONSKI *(continued)*

D. Thompson. Optical absorption in an electric field, *W. Liptap.* Electric dichroism of macromolecules, *C. Paulson, Jr.* Electric field light scattering, *B. Jennings.* Electrophoretic light scattering, *W. Flygare and B. Ware.* Circular dichroism and optical rotation in an electric field, *I. Tinaco, Jr.* Electro-optics in the infrared region, *E. Charney.* Electro-optic data acquisition and processing, *C. O'Konski and J. Jost.* Use of rotational diffusion coefficients for macromolecular dimensions and solvation, *P. Squire.* Electro-optics of polypeptides and proteins, *K. Yoshioka.* Electro-optics of polynucleotides and nucleic acids, *N. Stellwagen.* Electro-optics of polyectrolytes and dye-polyelectrolyte complexes, *M. Shirai.* Electro-optics of nucleoproteins and viruses, *M. Maestre.* Electro-optics of cells and membranes, *R. Keynes.* Electro-optics of liquid crystals, *T. Scheffer, H. Gruler, and G. Meier.* Non-linear electro-optics, *S. Kielich.* Magneto-electro-optics, *S. Kielich.* Quantum theory and calculation of electric polarizability, *T.-K. Ha.*

PARKS *Superconductivity*

In 2 Volumes

edited by RONALD D. PARKS, *University of Rochester, New York*

Vol. 1 688 pages, illustrated. 1969
Vol. 2 767 pages, illustrated. 1969

Reviews the subject of superconductivity in a truly comprehensive manner. Invaluable as a supplementary textbook for graduate students in physics and other scholars.

CONTENTS:

Volume 1: Early experiments and phenomenological theories, *B. S. Chandrasekhar.* Theory of Bardeen, Cooper, and Schrieffer, *G. Rickayzen.* Equilibrium properties: Comparison of experimental results with predictions of the BCS theory, *R. Meservey and B. B. Schwartz.* Nonequilibrium properties: Comparison of experimental results with predictions of the BCS theory, *D. M. Ginsburg and L. C. Hebel.* Green's function method, *V. Ambegaokar.* Ginzburg-Landau equations and their extensions, *N. R. Werthamer.* Collective modes in superconductors, *P. C. Martin.* Macroscopic quantum phenomena, *J. E. Mercereau.* Weakly coupled superconductors, *B. D. Josephson.* Electron-phonon interaction and strong-coupling superconductors, *D. J. Scalapino.* Tunneling and strong-coupling superconductivity, *W. L. McMillan and J. M. Rowell.* Superconductivity in low-carrier-density systems: Degenerate semiconductors, *M. L. Cohen.*

Volume 2: Superconductivity in transition metals: Theory and experiment, *G. Gladstone, M. A. Jensen, and J. R. Schrieffer.* Theory of type II superconductors, *A. L. Fetter and P. C. Hohenberg.* Type II superconductors: Experi-

ments, *B. Serin.* Boundary effects and small specimens, *J. P. Burger and D. Saint-James.* Proximity effects, *G. Deutscher and P. G. de Gennes.* Gapless superconductivity, *K. Maki.* Flux flow and irreversible effects, *Y. B. Kim and M. J. Stephen.* Comparison of properties of superconductors and superfluid helium, *W. F. Vinen.* Intermediate state in type I superconductors, *J. D. Livingston and W. DeSorbo.* Superconducting devices, *V. L. Newhouse.* Superconductivity in the past and future, *P. W. Anderson.*

PATRICK *Treatise on Adhesion and Adhesives*

edited by ROBERT L. PATRICK, *Alpha Research and Development, Inc., Blue Island, Illinois*

Vol. 1 *Theory*
496 pages, illustrated. 1967
Vol. 2 *Materials*
568 pages, illustrated. 1969
Vol. 3 *Special Topics*
264 pages, illustrated.

A compilation of existing theories, hypotheses, and postulates within the field of adhesion and its newer phenomenological aspects, as well as older emphasis on materials. The interdisciplinary contributions are of benefit to mechanical and chemical engineers, and to polymer and surface chemists.

CONTENTS:

Volume 1: Introduction, *R. L. Patrick.* Intermolecular and interatomic forces, *R. J. Good.* Adsorption of polymers, *R. R. Stromberg.* Mechanisms of adhesion, *J. R. Huntsberger.* Role of bulk properties of the adhesive, *T. Alfrey, Jr.* Rheology of polymers used as adhesives, *D. H. Kaelble.* Fracture mechanics applied to adhesive systems, *G. R. Irwin.* Variables and interpretation of some destructive cohesion and adhesion tests, *J. L. Gardon.* Surface chemistry, *F. M. Fowkes.*

Volume 2: Introduction, *R. L. Patrick.* Epoxide adhesives, *H. Dannenberg and C. A. May.* Thermosetting adhesives, *L. T. Eby and H. P. Brown.* Elastomeric adhesives, *W. C. Wake.* Pressure-sensitive adhesives, *C. A. Dahlquist.* Fiber adhesion, *H. T. Patterson.* Soldering, brazing, and welding. *F. H. Bair.* Glass resin adhesion in filament-wound structures, *S. Brelant.* High-temperature adhesion, *H. Levine.* Thermoplastic adhesives, *R. A. Weidener.* Preparation of ultraclean substrate surfaces, *V. Ponec.*

Volume 3: Structural adhesives for metal bonding, *J. Bolger.* Durability of adhesive bonded aluminum joints, *J. Minford.* Adhesion and the glassy state, *G. Miller.* The use of scanning electron microscopy, *R. Patrick.*

PODOLSKY and KUNZ   *Fundamentals of Electrodynamics*

by BORIS PODOLSKY and KAISER S. KUNZ, *New Mexico State University, Las Cruces.* Original text written by Boris Podolsky

512 pages, illustrated. 1969

A textbook for a one- or two- semester graduate or senior level course in electrodynamics or electromagnetic theory. Also of interest to physicists and electrical and electronic engineers.

CONTENTS: Introduction – general dynamics • Newtonian mechanics • The electrostatic field • Relativistic mechanics • The electromagnetic field • Radiation • Dynamics of charged particles • Material media.

SCHICK   *Nonionic Surfactants*

(Surfactant Science Series, Volume 1)
edited by MARTIN J. SCHICK, *Central Research Laboratories, Interchemical Corporation, Clifton, New Jersey*

1,120 pages, illustrated. 1967

CONTENTS: Introduction, *M. J. Schick.* **Part I: Organic Chemistry of Nonionic Surfactants:** Mechanism of ethylene oxide condensation, *N. Shachat and H. L. Greenwald.* Polyoxyethylene alkylphenols, *C. R. Enyeart.* Polyoxyethylene alcohols, *W. B. Satkowski, S. K. Huang, and R. L. Liss.* Polyoxyethylene esters of fatty acids, *W. B. Satkowski, S. K. Huang, and R. L. Liss.* Polyoxyethylene mercaptans, *H. Lemaire.* Polyoxyethylene alkylamines, *R. A. Reck.* Polyoxyethylene alkylamides, *E. Jungermann and D. Taber.* Polyol surfactants, *F. R. Benson.* Polyalkylene oxide block copolymers, *I. R. Schmolka.* Nonionics as ionic surfactant intermediates, *L. W. Burnette.* Miscellaneous nonionic surfactants, *L. W. Burnette.* Synthesis of homogeneous nonionic surfactants, *B. A. Mulley.* **Part II: Physical Chemistry of Nonionic Surfactants:** Surface films, *H. Lange.* Micelle formation in aqueous and nonaqueous solutions, *P. Becher.* Thermodynamics of micelle formation, *D. G. Hall and B. A. Pethica.* Solubilization, *T. Nakagawa.* Emulsification, *P. Becher.* Effect of nonionic surfactants on stability of dispersions, *R. H. Ottewill.* Detergency, *J. C. Harris.* Foaming, *G. M. Gantz.* Configuration of polyoxyethylene chain in bulk, *M. Rösch.* Configuration and hydrodynamic properties of polyoxyethylene chain in solution, *F. E. Bailey, Jr. and J. V. Koleske.* **Part III: Analytical Chemistry of Nonionic Surfactants:** Introduction to analytical chemistry of nonionic surfactants, *H. G. Nadeau and S. Siggia.* Noninstrumental methods of analysis. *H. G. Nadeau and S. Siggia,* Instrumental methods of analysis, *H. G. Nadeau and S. Siggia.* Separational methods, *H. G. Nadeau and P. H. Waszeciak.* **Part IV: Biology of Nonionic Surfactants:** Physiological activity of nonionic surfactants, *P. H. Elworthy and J. F. Treon.* Biodegradation, *M. J. Schick.*

SHINODA   *Solvent Properties of Surfactant Solutions*

(Surfactant Science Series, Volume 2)
edited by KŌZŌ SHINODA, *Department of Chemistry, Yokohama National University, Japan*

376 pages, illustrated. 1967

CONTENTS: Outline of solvent properties of surfactant solutions, *E. Hutchinson and K. Shinoda.* Solvent properties of nonionic surfactants in aqueous solutions, *K. Shinoda.* Interactions of polar molecules, micelles, and polymers in nonaqueous media, *F. M. Fowkes.* Physical chemistry of cleansing action, *H. Lange.* Pharmaceutical applications and physiological aspects of solubilization, *L. Sjöblom.* Surfactants in pesticidal formulations, *W. Van Valkenburg.* Emulsion polymerization, *B. M. E. van der Hoff.*

SWISHER   *Surfactant Biodegradation*

(Surfactant Science Series, Volume 3)
by R. D. SWISHER, *Monsanto Company, St. Louis, Missouri*

520 pages, illustrated. 1970

Shows the current state of knowledge of surfactant biodegradation as of mid-1969, and provides an extremely valuable point of departure for further work on other pollution problems. Of decided importance to workers in the fields of surfactants, detergents, and water pollution.

CONTENTS: Background and perspective • Surfactants – their nature, behavior, and structure • The analytical methods • The biological background • Biodegradation test methods • Chemical structure and primary biodegradation • Metabolic pathways and ultimate biodegradation • Biodegradation data.

VLADIMIROV   *Equations of Mathematical Physics*

(Pure and Applied Mathematics Series, Volume 3)
by VASILIY S. VLADIMIROV, *Steklov Institute of Mathematics, Moscow, U.S.S.R.* translated by AUDREY LITTLEWOOD translation edited by ALAN JEFFREY

428 pages, illustrated. 1971

Examines classical boundary value problems for differential equations of mathematical physics, using the concept of the generalized solution instead of the traditional means of presentation. Devotes a special chapter to the theory of generalized functions and may be used as a graduate text.

(continued)

VLADIMIROV *(continued)*

CONTENTS: Formulation of boundary value problems in mathematical physics • Generalized functions • Fundamental solutions and the Cauchy problem • Integral equations • Boundary value problems for elliptic equations • The mixed problem.

VÖLKEL Fields, Particles, and Currents

by A. H. Völkel, *Faculty of Physics, University of Bielefeld, West Germany*
in preparation. 1973

Provides an introduction to the theory of local quantized fields, gives the physical background ideas of the notion of a quantized field, and reviews the necessary mathematical tools. Contains the principal results on the structure of quantized fields which have been gained during the last 25 years. Of special benefit to students who intend to do research in this field. Also of use to physicists and mathematicians who want to increase their knowledge of the topic of local quantized fields.

CONTENTS: General notion of a quantized field • Free quantized fields • Local fields and Wightman distributions • Selected topics from the structure of local quantum theory • Spatial behavior of fields • Haag-Ruelle (Strong) asymptotic condition • Weak asymptotic conditions • Current generated charges and symmetrics • Field operators at short distances.

WELLER Solid State Physics and Chemistry: An Introduction

In 2 Volumes

edited by Paul F. Weller, *Department of Chemistry, State University of New York, Fredonia*

Vol. 1 576 pages, illustrated. 1973
Vol. 2 464 pages, illustrated. 1973

A text which presents the fundamental principles and practices of the solid state in a language and form that is understandable not only to those acquainted with solid state physics, but also to chemists and biologists. Establishes a firm foundation in solid state science for chemistry and biology students and any scientist interested in the field.

CONTENTS:
Volume 1: An introduction to principles of the solid state, *P. Weller.* Crystallography, *L. Suchow.* Bonding models of solids, *B. Crowder.* The electrical properties of solids, *J. Perlstein.* Magnetic properties, *J. Steger.* Introduction to magnetic resonance in solids, *P. Kasai.* Optical properties of solids, *J. Axe.*
Volume 2: Point defects in solids, *J. Prener.* Atom movements: Diffusion, *R. Brebrick.* Surface chemistry, *G. Somorjai.* Phase equilibria and materials: Preparation, *A. Reisman.* Crystal growth, *E. Kostiner.* Polymeric materials, *B. Wunderlich.* Biology and semiconduction, *H. Tien.*

ZETTLEMOYER Nucleation

edited by A. C. Zettlemoyer, *Lehigh University, Bethlehem, Pennsylvania*
624 pages, illustrated. 1969

Provides a source book and reference treatise for those interested in nucleation.

CONTENTS: General and theoretical introduction, *W. J. Dunning.* Homogeneous nucleation in a vapor, *R. P. Andres.* Statistical mechanics of nucleation, *J. Lothe and G. M. Pound.* Vapor to condensed–phase heterogeneous nucleation, *R. A. Sigsbee.* Nucleation in liquids and solutions, *A. G. Walton.* Nucleation of precipitates in defect solid solutions, *E. Hornbogen.* Condensation of metal vapors on substrates, *D. Walton.* Nucleation in polymer crystallization, *F. P. Price.* Nucleation in glass-forming materials, *J. J. Hammel.* Nucleation in the atmosphere, *E. A. Boucher.* Graining in sugar boiling, *A. VanHook.*

——— OTHER BOOKS OF INTEREST ———

MARINSKY *Ion Exchange: A Series of Advances*

edited by Jacob A. Marinsky, *Department of Chemistry, State University of New York, Buffalo*

Vol. 1 440 pages, illustrated. 1966
Vol. 2 256 pages, illustrated. 1969
Vol. 3 see Marinsky and Marcus for continuing volumes

MARINSKY and MARCUS *Ion Exchange and Solvent Extraction: A Series of Advances*

a series edited by Jacob A. Marinsky, *Department of Chemistry, State University of New York, Buffalo,* and Yizhak Marcus, *Department of Inorganic and Analytical Chemistry, The Hebrew University, Jerusalem, Israel*

for earlier volumes see MARINSKY
Vol. 3 168 pages, illustrated. 1973
Vol. 4 264 pages, illustrated. 1973
Vol. 5 256 pages, illustrated. 1973
Vol. 6 in preparation. 1974

MATTSON, MARK and McDONALD
Computer Fundamentals for Chemists

(Computers in Chemistry and Instrumentation Series, Volume 1)
edited by JAMES S. MATTSON, *Rosenstiel School of Marine and Atmospheric Sciences, University of Miami, Florida,* HARRY B. MARK, JR., *Department of Chemistry, University of Cincinnati, Ohio,* and HUBERT C. MACDONALD, JR., *Koppers Company, Inc., Monroeville, Pennsylvania*
384 pages, illustrated. 1973

MATTSON, MARK, and MacDONALD
Electrochemistry: *Calculations Simulation, and Instrumentation*

(Computers in Chemistry and Instrumentation Series, Volume 2)
edited by JAMES S. MATTSON, *Rosenstiel School of Marine and Atmospheric Sciences, University of Miami, Florida,* HARRY B. MARK, JR., *University of Cincinnati, Ohio,* and HUBERT C. MACDONALD, JR., *Koppers Company, Inc., Monroeville, Pennsylvania*
488 pages, illustrated. 1972

MATTSON, MARK, and MacDONALD
Spectroscopy and Kinetics
(Computers in Chemistry and Instrumentation Series, Volume 3)
edited by JAMES S. MATTSON, *Rosenstiel School of Marine and Atmospheric Sciences, University of Miami, Florida,* HARRY B. MARK, JR., *Department of Chemistry, University of Cincinnati, Ohio,* and HUBERT C. MACDONALD, JR., *Koppers Company, Inc., Monroeville, Pennsylvania*
352 pages, illustrated. 1973

MILLICH and CARREHER *Interfacial Synthesis*

edited by FRANK MILLICH, *Department of Chemistry, University of Missouri, Kansas City,* and CHARLES E. CARREHER,

JR., *Department of Chemistry, University of South Dakota, Vermillion*
in preparation. 1973

RIFI and COVITZ *Introduction to Organic Electrochemistry*

(Techniques and Applications of Organic Synthesis Series)
by M. R. RIFI and F. H. COVITZ, *Union Carbide Corporation, Bound Brook, New Jersey*
424 pages, illustrated. 1973

STEWART *Infrared Spectroscopy:*
Experimental Methods and Techniques

by JAMES E. STEWART, *Durrum Instrument Corporation, Palo Alto, California*
656 pages, illustrated. 1970

VIJH *Electrochemistry of Metals and Semiconductors*

(Monographs in Electroanalytical Chemistry and Electrochemistry Series)
by ASHOK K. VIJH, *Hydro-Quebec Institute of Research, Varennes, Quebec*
336 pages, illustrated. 1973

WALKER and THROWER *Chemistry and Physics of Carbon:* **A Series of Advances**

a series edited by PHILIP L. WALKER and PETER A. THROWER, *Department of Material Sciences, Pennsylvania State University, University Park*
Vol. 1 400 pages, illustrated. 1965
Vol. 2 400 pages, illustrated. 1966
Vol. 3 464 pages, illustrated. 1968
Vol. 4 416 pages, illustrated. 1968
Vol. 5 400 pages, illustrated. 1969
Vol. 6 368 pages, illustrated. 1970
Vol. 7 424 pages, illustrated. 1970
Vol. 8 480 pages, illustrated. 1973
Vol. 9 272 pages, illustrated. 1973
Vol. 10 288 pages, illustrated. 1973
Vol. 11 in preparation. 1973

WILSON *Radiation Chemistry of Monomers-Polymers-Plastics*

by JOSEPH E. WILSON, *Department of Chemistry, Bishop College, Dallas, Texas*
in preparation. 1974

———————JOURNALS OF INTEREST———————

JOURNAL OF MACROMOLECULAR SCIENCE—*Chemistry*

editor: GEORGE E. HAM, *White Plains, New York*

This international journal provides scientists with a cross-section of the outstanding contributions from laboratories around the world—published four and one-half months from publisher's receipt of last manuscript. The fields covered include anionic, cationic and free-radical addition polymerization and copolymerization, the manifold forms of condensation polymerization, polymer reactions, molecular weight studies, temperature-dependent properties, rheology, effects of radiation of all forms, polymer degradation, and many others.

8 issues per volume

JOURNAL OF MACROMOLECULAR SCIENCE—*Physics*

editor: PHILLIP H. GEIL, *Case Western Reserve University*

A periodical devoted to the publication of significant fundamental contributions concerning the physics of macromolecular solids and liquids. Papers deal with research in transition mechanisms and structure property relationships, the physics of polymer solutions and melts, of glassy and rubbery amorphous solids, and the physics of individual polymer molecules and natural polymers, as well as all the areas generally contained in polymer state physics. New instruments and techniques appropriate to polymer physics and short descriptions of new data are critically noted.

4 issues per volume

SEPARATION AND PURIFICATION METHODS

editors: EDMOND S. PERRY, *Eastman Kodak Company, Rochester, New York,* and CAREL J. VAN OSS, *State University of New York, Buffalo*

The purpose of this new journal is to cover all areas involving the separation and purification of both simple and complex compounds. Thus, articles deal with the separations of inorganic and organic substances, as well as biological materials. The techniques and methods discussed are of particular interest to biochemists, biophysicists, microbiologists, chemists, physicists, and chemical engineers. *Separation and Purification Methods* provides authoritative summaries on significant new developments and critical evaluations of new methods, apparatus, and techniques.

2 issues per volume

SEPARATION SCIENCE

An Interdisciplinary Journal of Methods and Underlying Processes

executive editor: J. CALVIN GIDDINGS, *University of Utah, Salt Lake City*

This is a fundamental journal whose articles probe the very essence of separation phenomena, and consequently evolve new concepts and techniques for dealing with the problems of this enormously important field. It contains authoritative and critical articles, notes, and reviews dealing wih all scientific aspects of separations and will unify this greatly fragmented field. It also stimulates new developments through cross-fertilization and enhances and supports the efforts of workers in biology, chemistry, engineering, and other sciences.

6 issues per volume

SPECTROSCOPY LETTERS
An International Journal for Rapid Communication

executive editor: JAMES W. ROBINSON, *Louisiana State University, Baton Rouge*

This journal provides a rapid means of publication of fundamental developments in spectroscopy. Fields included are NMR, ESR, microwave, IR, Raman, UV, atomic emission, atomic absorption, X-ray, nuclear science, mass spectrometry, lasers, electron microscopy, molecular fluorescence, and molecular phosphorescence.

12 issues per volume

TRANSPORT THEORY AND STATISTICAL PHYSICS
An International Journal for Rapid Communication in Irreversible Statistical Mechanics

editor: PAUL ZWEIFEL, *Virginia Polytechnic Institute, Blacksburg*

This new journal will be devoted primarily to those areas of irreversible statistical mechanics which are generally known as transport theory and kinetic theory. It will include papers on transport of neutral particles, dynamics of simple liquids, kinetic theory of liquids and plasmas, gas dynamics, and correlation functions.

4 issues per volume

———— ENCYCLOPEDIAS OF INTEREST ————

ENCYCLOPEDIA OF THE ELECTROCHEMISTRY OF THE ELEMENTS

editor: ALLEN J. BARD, *University of Texas at Austin*

This new multi-volume *Encyclopedia* presents a critical review and a compilation of the descriptive electrochemistry of the elements and their compounds. Each chapter includes tables of available thermodynamic and kinetic data, a listing of known electrochemical reactions, and a discussion of the mechanism of these reactions, when known, for the element in question and its compounds. A notable feature of the *Encyclopedia* is a section on the applied electrochemistry of the elements and their compounds, and their use in electrochemical devices. Also included are references to the original literature on all data and reactions. This compilation provides a starting point for new electrochemical investigations and suggests the areas in which further research is necessary. Periodic supplements are planned in order to make available to the reader the most current information in the field.

Encyclopedia of the Electrochemistry of the Elements will be of utmost interest to graduate students and researchers in electrochemistry, analytical, physical, inorganic and organic chemistry, chemical engineering, and corrosion science.

VOLUME 1 — CONTENTS

Chlorine, *T. Mussini and G. Faita.* Bromine, *T. Mussini and G. Faita.* Iodine and Astatine, *P. Desideri, L. Lepri, and D. Haimler.* Cadmium, *N. Hampson and R. Latham.* Lead, *T. Sharpe.* Manganese, *C. Liang.* Calcium, Strontium, Barium, and Radium, *S. Toshima.* Inert Gases, *B. Jaselskis and R. Krueger.*